Governing Childhood into the 21st Century

CRITICAL CULTURAL STUDIES OF CHILDHOOD

Series Editors:
Marianne N. Bloch, Gaile Sloan Cannella, and Beth Blue Swadener

This series will focus on reframings of theory, research, policy, and pedagogies in childhood. A critical cultural study of childhood is one that offers a "prism" of possibilities for writing about power and its relationship to the cultural constructions of childhood, family, and education in broad societal, local, and global contexts. Books in the series will open up new spaces for dialogue and reconceptualization based on critical theoretical and methodological framings, including critical pedagogy, advocacy and social justice perspectives, cultural, historical and comparative studies of childhood, post-structural, postcolonial, and/or feminist studies of childhood, family, and education. The intent of the series is to examine the relations between power, language, and what is taken as normal/abnormal, good and natural, to understand the construction of the "other," difference and inclusions/exclusions that are embedded in current notions of childhood, family, educational reforms, policies, and the practices of schooling. *Critical Cultural Studies of Childhood* will open up dialogue about new possibilities for action and research.

Single authored as well as edited volumes focusing on critical studies of childhood from a variety of disciplinary and theoretical perspectives are included in the series. A particular focus is in a re-imagining as well as critical reflection on policy and practice in early childhood, primary, and elementary education. It is the series intent to open up new spaces for reconceptualizing theories and traditions of research, policies, cultural reasonings and practices at all of these levels, in the USA, as well as comparatively.

The Child in the World/The World in the Child: Education and the Configuration of a Universal, Modern, and Globalized Childhood
   Edited by Marianne N. Bloch, Devorah Kennedy, Theodora Lightfoot, and
   Dar Weyenberg; Foreword by Thomas S. Popkewitz

Beyond Pedagogies of Exclusion in Diverse Childhood Contexts: Transnational Challenges
   Edited by Soula Mitakidou, Evangelia Tressou, Beth Blue Swadener, and Carl A. Grant

"Race" and Early Childhood Education: An International Approach to Identity, Politics, and Pedagogy
   Edited by Glenda Mac Naughton and Karina Davis

Governing Childhood into the 21st Century: Biopolitical Technologies of Childhood Management and Education
   By Majia Holmer Nadesan

Developmentalism in Early Childhood and Middle Grades Education: Critical Conversations on Readiness and Responsiveness
   Edited by Kyunghwa Lee and Mark D. Vagle

GOVERNING CHILDHOOD INTO THE 21ST CENTURY

Biopolitical Technologies of Childhood Management and Education

Majia Holmer Nadesan

GOVERNING CHILDHOOD INTO THE 21ST CENTURY
Copyright © Majia Holmer Nadesan, 2010.

First published in 2010 by
PALGRAVE MACMILLAN®
in the United States—a division of St. Martin's Press LLC,
175 Fifth Avenue, New York, NY 10010.

Where this book is distributed in the UK, Europe and the rest of the
world, this is by Palgrave Macmillan, a division of Macmillan Publishers
Limited, registered in England, company number 785998, of Houndmills,
Basingstoke, Hampshire RG21 6XS.

Palgrave Macmillan is the global academic imprint of the above companies and has companies and representatives throughout the world.

Palgrave® and Macmillan® are registered trademarks in the United States,
the United Kingdom, Europe and other countries.

ISBN: 978–0–230–61321–8

Library of Congress Cataloging-in-Publication Data is available from the
Library of Congress.

A catalogue record of the book is available from the British Library.

Design by Newgen Imaging Systems (P) Ltd., Chennai, India.

First edition: April 2010

10 9 8 7 6 5 4 3 2 1

Printed in the United States of America.

This book is dedicated to my husband, Alexander Govind Nadesan,
and to children everywhere.

Contents

CHAPTER 1

Introduction to Biopolitics, Risk, and Childhood

Across time, children in Western cultures have been understood as threatening beings aligned with satanic influences, as miniature adults, as fragile creatures of God, as delinquents, and as vulnerable, at-risk beings. In exploring these historically specific formulations of childhood, academic and cultural observers have drawn attention to the complex ways in which social identities are forged in time through institutional positionings, technologies of government, risk-communication systems, and everyday practices. Analyses of childhood reveal that symbolic formulations and preferred and denigrated strategies of raising children point to changing socioeconomic problematics and strategies of government. Childhood is not a natural space but rather is carved out by culturally and historically specific technologies of government.

This project continues efforts to understand the social and historical constitution of childhood and children using governmentality as a framework of analysis. Governmentality addresses how society's pressing problems, expert authorities, explanations, and technologies are organized in relation to particular kinds of problem-solution frames (see Rose *Governing* xi). Governmental analysis derives from the work of Michel Foucault (1926–1984), but with a diverse and interdisciplinary group of scholars appropriating the framework to study those logics of government and problem-solution frames that shape the conduct of everyday practices, it has since proliferated in forms and applications. Foucault-inspired governmental analyses of childhood have been pioneered by Philippe Ariès, Jacques Donzelot, Nikolas Rose, and Michael Peters, among others, and have more recently informed studies of the sociology of childhood, education, and childhood and popular culture.[1]

This book extends this disparate governmental scholarship on childhood by focusing on how childhood has been shaped by biopolitical strategies of risk management. The concept of "risk" increasingly dominates popular discourse, public policy, and marketplace problem-solution frames. Richard Ericson and Aaron Doyle define risk as "threats or dangers attributed to persons, technologies, or nature" (2). Accordingly, Ulrich Beck asserts: "Being at risk is the way of being and ruling in the world of modernity; being at global risk is the human condition at the beginning of the twenty-first century" (330). Financial management, social services, policing, development policy, national defense, and environmental policy are just a few of the areas that have come to rely on common vocabularies of risk (Baker and Simon 1). Although risk analysis can tend toward political or technological realism, this book explores how particular formulations of risk operate as governmental rationalities by shaping perceptions, problem-solution frames, and action orientations.

In particular, this project focuses on biopolitical technologies of risk management shaping expert, market, and everyday understandings of children and practices of child rearing historically and in twentieth- and early twenty-first-century America. Foucault developed the idea of biopolitics to capture how technologies of power help in managing, and also controlling, the life of the population. Although biopolitical knowledge and practices often derive from expert understandings—from knowledge formations produced by educational, psychological, and psychiatric authorities—they also derive from economic authorities and from everyday people engaged in everyday routines and practices.

In the early twentieth century, philanthropists and the emerging authorities of the helping professions (e.g., social workers and psychologists) offered historically novel formulations of childhood as "at risk," thereby demanding expert oversight and guidance. These formulations encouraged expansion of social welfare institutions designed to produce normalized, economically integrated adults out of malleable, vulnerable, or dangerous children. During the 1930s, social welfare strategies of government were legitimized by economic authorities such as John Maynard Keynes (1883–1946). Other economic authorities, such as F. A. Hayek (1899–1992), condemned the expansion of collectivized risk-management strategies by arguing that they encouraged dependency. Everyday people, such as parents, actively sought out or resisted these social welfare institutions and expert authorities, depending upon their circumstances and inclinations.

By the mid-twentieth century, middle-class and upper-class childhoods were understood as fraught with educational, cultural, and environmental risks that required careful parental oversight from early infancy onward. Children were at risk from too much television, from inadequate academic preparation, from lack of exercise, and so on. Concerned parents and marketing authorities looked to people with educational and psychological expertise when developing and implementing strategies, technologies, and products for navigating childhood perils. With growing awareness of vulnerability to environmental insults such as poisoning caused by lead, childhood became ever more risky and required further protections enacted by the state and by vigilant parents.

Yet despite growing concerns about risks to childhood, children of the lower socioeconomic classes were more likely to be regarded as "risky" than "at risk" throughout most of the twentieth century. Early twentieth-century child saving and child guidance had, after all, been prompted by the specter of the "dangerous" child of the nineteenth century. The establishment of social welfare institutions and the logics of government in the 1930s and their extension in the 1960s helped soften, but did not eliminate, negative public attitudes toward lower-class children. White, popular sentiment often regarded poor children, particularly minority poor children, as inherently deficient and therefore fundamentally risky. Eugenic formulations of the early twentieth century that explained dangerousness in relation to heritable degeneracy simply ceded to culturalist accounts of intergenerational poverty. At the close of the twentieth century, heredity accounts again gained currency in explaining children's inability to pull themselves up by the proverbial bootstraps.

At the dawn of the twenty-first century, the influence of traditional, twentieth-century childhood authorities (e.g., school authorities and pediatricians) waned as middle-class and upper-class parents sought advice for navigating risks to optimal childhood development from new information sources (e.g., Internet sites), fragmenting the array of forms of childhood expertise. The fragmentation of authority over childhood and the growth of a marketplace of child-related goods produced a seemingly diverse array of discourses aimed at describing, normalizing, disciplining, optimizing, and celebrating childhood. However, this apparent diversity in childhood formulations masks an increasing tendency to regard middle- and upper-class childhood as fraught with biopolitical risks. Children are at risk from spending too much time in front of the computer, from obesity, from underachieving schools, and from environmental toxins in plastic bottles, among other proliferating risks.

In the white, middle-class popular imagination, contemporary American children are also constituted as at risk from dangerous foreign or ethnic cultures and from the indolent tendencies of peers whose families fail to practice the entrepreneurial technologies of the self necessary for future success. In effect, white, suburban, middle-class children are perceived as at risk from being exposed to risky children. Risky children exist domestically and abroad. Risky, domestic children are future welfare dependents or future criminals whose environmental or biologically mediated riskiness escapes remediation by the helping professions. Risky children abroad threaten to reject American values and the global, neoliberal economic order. Or, conversely, risky children abroad threaten the "American way of life" by their very success within the global economic order, by outperforming American children in math and technology, and by their future capacity to work more cheaply than Americans.

Particularly interesting and important about these constitutions of "risky" children are the technologies of government that are being proposed and adopted for their management. Throughout much of the mid-twentieth century, social welfare logics, technologies, and strategies of government shaped representations of childhood risks and informed strategies for their management. However, neoliberal logics and technologies have supplemented or replaced social welfare ones over the last thirty years in the United States and abroad. Foucault describes neoliberalism in *The Birth of Biopolitics* in terms of the prioritization and extension of economic logics: "The generalization of the economic form of the market beyond monetary exchanges functions in American neo-liberalism as a principle of intelligibility and a principle of decipherment of social relationships and individual behavior" (243). Neoliberalism shapes domestic life, in that market logics and technologies of the self define attitudes toward children, child rearing, and education. Neoliberalism shapes international life, in that market-influenced biopolitical strategies and technologies of government shape international relationships, foreign aid, philanthropy, and military planning.

This project explores these neoliberal representations and technologies in terms of their formulations and management of biopolitical risks for, and from, American children. Additionally, this project also explores how neoliberal biopolitical problem-solution frames represent and seek to govern the childhoods of other nations by regulating children's transnational flows and by shaping the conditions of their existence through new and old security apparatuses. The increasing tendency to represent foreign children in terms of their "security" risks for the United States has ominous implications in the context of the current economic crisis and in relation to the social upheavals caused by climate change, desertification, depleted water supplies, and conflict over scarce resources.

On the one hand, as illustrated above, the tendency to consider and evaluate children and childhood within risk frames of reference is not new. Children have been considered both at risk and risky since child savers and social Darwinists vied for the power to define childhood at the close of the nineteenth century. On the other hand, contemporary neoliberal logics of economic government and technologies of the self (Foucault *Biopolitics* 243–48), coupled with epidemiological models of public safety (Rose *Politics* 226) and targeted governance (Valverde and Mopas 232–50), converge to escalate the social salience and urgency of risk in biopolitical formulations of childhood. This project addresses how neoliberalism is driving the increased salience of risk in understanding and governing childhood and examines how neoliberal problem-solution frames are replacing social welfare ones.

Some readers may wonder whether the election of the U.S. president Barack Obama in 2008 and the political fallout from the financial crisis that began in late 2007 might have dislodged neoliberal logics of government. The position developed in this project holds that current U.S. government and market policies and practices continue to elevate the market as the primary technology for distributing societal resources and for evaluating the distribution of societal resources. The extensive U.S. government bailouts of market actors, including commercial (e.g., Citibank) and investment financial institutions (e.g., Goldman Sachs), illustrate the subordination of (formal U.S.) government to the financial industry. Indeed, a 2009 article in the *Atlantic Monthly* titled "The Quiet Coup" argues that the U.S. "financial industry has effectively captured our government" (S. Johnson 46). The subordination of government policy, priorities, and funding to neoliberal financial interests and markets points to a retrenchment of neoliberalism, rather than to its dissolution. The final chapters of this book will explore the implications of this development for future biopolitical formulations of U.S. childhood.

Accordingly, this project updates and extends Foucauldian and sociological studies of childhood in several ways. First, it incorporates recent work studying the biopolitics of childhood by addressing how risks to children's optimization are constructed, represented, and governed biotechnologically and culturally through a wide array of expert and everyday cultural techniques. Second, it incorporates political economy by addressing how neoliberal economic and cultural technologies of government produce the conditions of possibility for this intensive biopolitics of childhood, shaping risk-management technologies in accord with economic logics and status-based opportunities. Third, it addresses how U.S. media, government policy, and security discourses seek to govern global flows of children and childhood development abroad. Finally, it concludes by considering the future of childhood in the United States in the context of the momentous economic events of the U.S. financial meltdown.

In what follows, the reader is introduced to the theoretical framework of governmentality used to integrate the diverse topics pertaining to childhood examined in this project. This chapter concludes by outlining the book's organization. The reader will find it possible to peruse the chapters sequentially or according to their particular interests. Chapters 3 through 6 deliberately place theoretical considerations in the background, excepting concluding remarks. Readers primarily interested in an empirical account of the history of the present may simply skip this introduction to the theoretical grounds of this project.

GOVERNMENTALITY: A BRIEF INTRODUCTION

Foucault offers readers a historical genealogy of Western, liberal logics of government and technologies of power from the eighteenth century onward in his classic essay "Governmentality" and more recently in *Security, Territory, Population* and *The Birth of Biopolitics.* This genealogy embraces several historical milieus Foucault delineates based on their distinct problem-solution frames and governmental logics, including state mercantilism, laissez-faire liberalism, social welfare liberalism, and neoliberalism. Problems and logics of government and technologies of power shape the characteristic institutions, practices, authorities, and subjectivities of these social milieus. Technologies of power include (a) "technologies of production"; (b) "technologies of sign systems"; (c) "technologies of power"; and (d) "technologies of the self" (Foucault *Technologies* 18). Risk, as constructed through sign systems, materialized in policy, and performed through technologies of the self, often (but not always) accords with shifting liberal logics and technologies of government.

State Mercantilism and Laissez-Faire Logics of Government

The logic of government producing that milieu of power characterized as state mercantilism during the late 1600s and 1700s emphasized the reason of the state and the forces necessary for enhancing the wealth and power of the state. Foucault describes how the idea of the territorially delimited state became a "principle of intelligibility" after the Treaty of Westphalia of 1648 (*Security* 286). Efforts to rationalize the state's existence outside of the divine right of kings led to increased formalization of state apparatuses of government and heightened efforts to accumulate national wealth. Wealth was no longer tied to the land but was tied to production, linking it closely to the population. Efforts to document and accumulate wealth through the population's productive capacities led to the creation of new authorities— political statisticians—who used statistics to establish links across the health, productivity, and wealth of nations. Therefore, the early reason of the state that guided sovereign authorities was based on a "continuous and multiple network of relationships between the population, the territory, and wealth" (*Security* 106). The sovereign was responsible for increasing all of the forces of the state from within, including the tightly coupled relationship of population and wealth, while developing military forces and diplomatic relations capable of protecting the territorially delimited nation.

Foucault describes *police* as the science of administering the population that began under state mercantilism but grew in significance under nineteenth-century laissez-faire liberalism (*Biopolitics* 7). The apparatuses of police, such as public sanitarians and constables, tried to securitize the population from risks to their health and economic productivity, including epidemics, death in infancy, and vagrant populations who threatened evolving moral sensibilities and failed to conform to the growing demands of economic industrialization. The logic of police entailed a kind of secularized Christian ecclesiastic authority, based on the model of the shepherd's relation to his flock. The more pastoral model of police was supplemented with the ancient sovereign power to kill, to torture, to incarcerate, and to deny the conditions of life through decree.

Extension of the science and administration of police coincided with less direct monarchical sovereignty over the economy as laissez-faire, liberal economic logics gained in social importance. Liberalism, as that sign system articulated in the philosophies of John Locke (1632–1704) and Adam Smith (1723–1790), seized upon the market as a test case for limiting sovereign control (*Biopolitics* 45–56, 320). Foucault observes that the market's relevance as test stemmed from the "fundamental incompatibility between the optimal development of its economic process and a maximization of governmental procedures" (320). The early liberal critique of sovereign power helped free "reflection on economic practices from the hegemony of the '*raison d'État*'" (320). Thus, a primary problem-solution frame for laissez-faire liberalism involved freeing market circulations from direct sovereign control. For example, early liberals wanted the growing markets of stocks in

Amsterdam and London to operate freely without state restraint. At the same time, however, merchants found the existing legal infrastructures insufficient for protecting newly emerging forms of property and therefore pressed for the extension of the legal-juridical forms of state government that could recognize and protect bourgeois property. Hence, the market emerged as a space that was both at risk from (excessive) state control and in need of the state's infrastructural, particularly juridical, supports.

The technologies of government that came to be regarded as laissez-faire liberalism by the early 1800s enacted symbolic and material distinctions among socially delineated spheres of market, population, and state. Laissez-faire liberal philosophy presumed that each sphere was capable of self-regulation. The market was presumed to be governed by the guiding hand. The populace, or population, was presumed to be governed by the individual pursuit of liberty and market self-interest, tempered by prudence, and shepherded by the pastoral apparatuses of police. This formulation of the populace presumed a model of subjectivity Foucault labels *homo oeconomicus* (*Biopolitics* 270–71). Homo oeconomicus emerged as the "intangible partner of *laissez-faire*" (270). Laissez-faire philosophy held that rational, self-interested market behavior produced positive collective outcomes for society, thereby aligning private market interests with public advantage (A. Smith 572). In principle, homo oeconomicus curtails state power, as it "strips the sovereign of power inasmuch as he reveals an essential, fundamental, and major incapacity of the sovereign, that is to say, an inability to master the totality of the economic field" (*Biopolitics* 292). The state's legitimacy had to derive from new functions, given its loss of authority over the market. Accordingly, the state sought to promote the fecundity of the population and to arbitrate among potentially competing interests. Thus, the state increasingly formalized its structure in juridical codes and laws that carved up and distinguished among political, economic, and private spaces in terms of rights and obligations, thereby creating that space described as "civil society" (295–96).[2] The state's governmentality of civil society was limited by "the rights of man," a formulation of liberty seen as complementing homo oeconomicus (328). The incitements to liberty promoted by the rights of man and homo oeconomicus led the state to create security mechanisms to hedge and structure risks stemming from liberal freedoms.

Liberal governmentalities are organized around "apparatuses of security" such as standing armies, police forces, diplomatic corps, intelligence services, and social welfare services, including health and education systems (Dean *Governmentality* 20). Security apparatuses promote the strength of the nation and protect its population from threats, both internal and external. Security mechanisms entail and require biopolitical knowledge of the species body. In the *History of Sexuality*, volume 1, Foucault explains that biopolitics addresses the species body of the population more generally. Its concerns embrace "the body imbued with the mechanics of life and serving as the basis of the biological processes: propagation, births and mortality, the level of health, life expectancy and longevity, with all the conditions that can cause these to vary" (139). Biopolitics thus involves "distributing the living in the domain of value and utility" and has the power "to qualify, measure, appraise, and hierarchize," ultimately producing "distributions around the norm" (144). Foucault concludes that a "normalizing society is the historical outcome of a technology of power centered on life" (144). Children emerge as central targets of these normalizing forces in the nineteenth century.

In *The Order of Things*, Foucault describes how late nineteenth-century academic disciplines and professional practices of psychology established biopolitical norms that served as the basis for new technologies of power that operated upon the population to elicit and channel its vital forces while reducing risks to those forces (384–86). Institutional observations and case-by-case analyses provided by doctors, psychiatrists, criminologists, and so on, played an important role in developing these norms. As Foucault discusses in *Discipline and Punish,* these norms were then used to discipline individuals in enclosed institutional spaces such as schools, hospitals, and factories. While many of these disciplines simply aimed at exhorting bodily energies in structured manners, others sought to cultivate healthier bodies through sanitation techniques such as hand washing. Popular dissemination of sanitation practices in the mid-nineteenth century in magazines and pamphlets directed at the public encouraged willing citizens to adopt technologies that promised to foster their health and happiness (Tomes 54). In the late nineteenth century, popular periodicals and the new agents of consumer culture disseminated psychological, criminological, medical, and developmental norms being created by the growing cadres of authorities of the human sciences. Taken as a whole, the new biopolitical disciplines, norms, and technologies of the self both acted upon individuals as objects within a mass and were aimed at willing individuals, thereby operating as techniques of subjectification by transforming individuals into particular kinds of subjects who reflected upon and acted upon themselves in accord with newly established norms of health, manners, conduct, sociality, and affect (Foucault "The Subject and Power" 208).

These pastoral dynamics were, and continue to be, supplemented with repressive operations executed by security apparatuses. Individuals who fail to conform to liberal formulations of personhood have been subjected in the past and present to outright extermination, enclosure, disciplinary normalization, and, in the twentieth century, to circulating and preventive "biopolitical" security apparatuses. Chapter 2 provides a historical account of how various types of security apparatuses have targeted, constituted, and disciplined childhood. The rest of this project investigates how biopolitical distinctions are linked to various types of security strategies including force, discipline, normalization, and optimization. The strategies of normalization and optimization are historically linked to social welfare logics of government.

Social Welfare Government, Norms of Health and Illness, and Epidemiological Medicine

As expressed by Dr. W. Ford at the 1915 meeting of the Association of American Physicians, in the early years of the twentieth century there was an

> insistent demand...for better facilities for the training of public health officials. This demand has come from members of the medical profession, chiefly those engaged in official positions as officers or commissioners of health for cities and states, from sanitary engineers, and from various philanthropic societies whose aim is the betterment of social conditions among the poor in our great cities and in our rural communities. With the last this demand is associated with a demand for more enlightened instruction for the general public in matters affecting their health. (1)

According to Ford, hygiene requires efforts by experts to clearly "formulate the underlying principles of this science, its scope and its needs" (1). Epidemiology would become an important representational technology of that branch of science then known as *hygiene*, which operated as an important security technology in the early twentieth century.

In the 1920s the field and science of hygiene would be renamed "public health," better signifying the totality of its embrace (Winslow 23–33). The imperatives of public health would usher in new social welfare policies, thereby expanding social welfare logics of government. The knowledge and practices of epidemiology and inoculation/vaccination served together as the vehicles for rationalizing the extension of social welfare logics. To prevent disease outbreaks, epidemiologists surveyed populations for sites of likely irruption. The relationship between poverty and disease discovered through careful surveillance suggested possibilities for preventive hygiene through improved sanitation and nutrition. Criminality, madness, and juvenile delinquency were soon subject to the same epidemiological approach, paving the way for the mental hygiene movement. The concept of dangerousness would cede to the concept of risk in new epidemiological representations that helped shift blame away from "defective" individuals toward deleterious environmental conditions. A new hygienic orientation targeted children's welfare and women's fecundity in particular and would give rise to social welfare apparatuses that in some Western nations profoundly altered, replaced, or supplemented laissez-faire liberal logics and problems of government.

The "problem" of child delinquency illustrates how epidemiological data and preventive eugenics rationalized and facilitated the expansion of social welfare logics. Childhood delinquency was discovered as a problem in the nineteenth century (see Platt 3). This discovery occurred in the context of concerns about race suicide as white European populations looked with dismay upon demographic results that showed declining birthrates among middle-class bourgeois populations. In April 1912 the United States created the Children's Bureau, which functioned as a national center of information, research, and education "as to the needs of the whole child and the interrelated problems of health, dependency, delinquency and the employment of children" (Abbott XX4). Twenty years later, in 1932, Grace Abbott, then the head of the Children's Bureau, articulated its accomplishments. Her account is noteworthy both for its use of epidemiological methods to illustrate problems and for its illustration of the social welfare reforms instituted in the name of the child during the early decades of the twentieth century.

Accordingly, Abbott observes in 1912 that the "birth registration area" had not been established, so officials lacked knowledge about infant births and mortality, although delinquency rates were slowly being tabulated by "experts" such as William Healy, who opened the U.S. Juvenile Psychopathic Institute in 1909 (XX4). By 1932 the birth registration area comprised forty-six states and the District of Columbia. Estimates existed only for the infant mortality rate of 1912, Abbott explains, but in 1930 approximately 150,000 infants died, which, she notes, compared favorably with the estimated 300,000 deaths thought to have occurred in 1912. She also notes approvingly that rates of child employment declined significantly across the period ranging from 1912 to 1930. Finally, Abbott claims that childhood delinquency rates declined as well.

Abbott attributes the decline in infant mortality, childhood labor, and childhood delinquency to several important government-supported social welfare programs. She cites the expansion of outdoor relief in the form of mothers' pension laws in forty-five states and the District of Columbia as reducing dependency and child labor. She also cites with approval the establishment of child welfare bureaus in more than half the states, which "promote the interest of dependent, neglected and delinquent children" (XX4).

She particularly emphasizes the role of the juvenile justice system in improving childhood conditions. All but two states, she explains, have juvenile court laws, and many of these courts are aimed at "cure" rather than punishment. Accordingly, she writes that although the juvenile court system was under way in 1912, it remained grounded in the nineteenth-century tradition of punishment. In contrast, "Under the new legal theory of the juvenile court children are all treated alike only when they are all treated differently. The question is not what should be done for particular types of offenses, but what should be done for individual children. The objective is cure, not punishment" (XX4). Abbott notes that this approach is in practice difficult to administer and requires extensive knowledge of childhood differences:

> The attempt to *prevent or cure delinquency* among children has led to the study of physical and mental defects of children, to investigation of their mental and personality problems, to examination of their environments for causes of maladjustment and experiment in individual or group training during the pre-school period. The first psychiatric clinic for the study of child delinquency, established largely through the efforts of Julia Lathrop, was still regarded as experimental in 1912. There are more than 600 psychiatric and child guidance clinics at the present time. (XX4)

In effect, efforts to prevent and cure delinquency led to the creation of new public apparatuses, including the juvenile courts and special schools for truants; new specialized professional authorities, such as probation officers, physicians, and psychiatrists attached to juvenile courts; and new spaces, such as the child psychiatric clinic and the child guidance clinic. New spaces created opportunities for the new authorities to generate new biopolitical norms of childhood (see E. Murphy 472–76).

Finally, Abbott stresses how expanded apparatuses of public health improved child well- being, largely through educating parents. She points out that in 1912, only one U.S. state had a bureau of child health, whereas in 1932, one-sixth of the counties in the nation had adopted organized county plans for health or social services or both. She views the "child health centre" as "the great teaching agency" largely responsible for improved child welfare. Moreover, unlike the child health specialists of the past, specialists of the 1930s devoted as much as one-third of their time to the supervision of "well children" in order to better understand childhood norms. Abbott praises how specialist knowledge of normal and deviant development was made available to parents: "Less expensive and more practical books giving the general principles of child care are now available and widely used" (XX4). She laments that the demand for popular bulletins exceeds the appropriations budget of the U.S. Children's Bureau. Abbott concludes that the parents of 1932 "know more about the scientific care of children than did those of 1912 and they are using more intelligently the knowledge and skill of the physician" (XX4).

Abbott's essay illustrates how efforts to collect epidemiological data in the early twentieth century on public concerns such as infant mortality, child labor, and delinquency led to the institutionalization of a nexus of public and private authorities, spaces, and specialized knowledges and treatment protocols all organized around saving children and preventing delinquency. These spaces, authorities, knowledges, and protocols were security mechanisms that operated to enhance the vitality and economic productivity of the nation. While these security mechanisms were first associated with disciplinary spaces such as the criminal courts and, later, the juvenile justice system, they also became disembedded from particular locales and circulated throughout everyday life in the form of the pamphlets and advice columns targeting parents who wished to prevent their child's future delinquency. Katherine Jones's *Taming the Troublesome Child: American Families, Child Guidance, and the Limits of Psychiatric Authority* and Theresa Richardson's *The Century of the Child: The Mental Hygiene Movement and Social Policy in the United States and Canada* provide the interested reader with careful histories of how child saving became an everyday concern of child rearing.

In sum, the birth and growth of epidemiology, its attendant mantra of prevention, and its historical applications to child saving helped destabilize laissez-faire logics and problems of government. Risk was socialized in the twentieth century with the institutionalization of state apparatuses that adapted the social model of medicine and public health to promote national security, cast in biopolitical terms. The degree to which risk was socialized and disseminated among the population in the form of social security technologies for reducing personal risks posed by unemployment, retirement, health, and education varied from nation to nation.

However, despite differences in form and degree of societal penetration, social welfare apparatuses tended to use a common logic of government based in an "actuarial model" for measuring and redressing risk (O'Malley *Risk* 18). Actuarial models tend to dissolve concrete subjects (i.e., individuals) in favor of a focus on seemingly objective, statistically derived risk factors dispersed across populations. Actuarial models are, as described by Pat O'Malley, "aggregating," since they locate individuals in groups based on specifically delineated characteristics (*Risk* 18). These characteristics are typically delineated in relation to risks such as those linked with drug abuse, alcoholism, criminality, and so on. Individuals grouped by shared riskiness can then be targeted for heightened surveillance or governance. Actuarial models can deconstruct the individual as a distinct subject because the field of statistical visibility entails the compilation and combinations of "factors" likely to produce biological, environmental, behavioral, and mental risks (Castel 287). As Castel explains:

> A risk does not arise from the presence of particular precise danger embodied in a concrete individual or group. It is the effect of a combination of abstract *factors* which render more or less probable the occurrence of undesirable modes of behavior [or illness, mental illness, dependency, etc.]. (287)

Whereas risk in the nineteenth century resided in concrete and often "dangerous" places and individuals, risk in the late twentieth century was constituted at the level of the population. Elaborating on Castel's formulation of risk, Paul Rabinow explains how twentieth-century surveillance strategies project (socially) determined risk factors onto the population: "This new mode anticipates possible loci of dangerous irruptions, through the identification of sites statistically locatable in relation to norms and means" (187). Computer technology eventually facilitated this decontextualized, impersonal assessment of at-risk and/or risky populations and their economic costs at the close of the twentieth century.

From the 1930s through the 1970s, social welfare authorities' efforts to explain heightened risk tended to problematize social phenomena such as poverty, family dynamics, education, and so on. Keynesian-inspired economic and education policies sought to redress "social problems" believed to explain heightened risk.[3] Social welfare liberalism encouraged the state to assume greater responsibility for managing and securing health and environmental (social and ecological) risks to the population. Thus, state-sponsored biopolitical authorities anticipated risks and acted to avoid or manage their detrimental effects, while simultaneously helping to cultivate the health of the population through education and health-promotion programs. In the United States, social welfare health programs remain firmly entrenched within contemporary life in the institutional edifices of the Centers for Disease Control, the National Institutes of Health, Medicare, and state-sponsored health-welfare programs. The agents and products of the expansive consumer culture introduced still other practices and products that promised to ensure happiness, health, and economic success.

Neoliberal Governmentalities

Within the last thirty years, many of the social welfare logics and problematics of government have been subject to revision or have been replaced by neoliberal, neoconservative, and Christian pastoral logics and problematics (see Nadesan *Governmentality*). Neoliberalism provides an economic logic and approach toward risk management shared by the other two approaches, so discussion focuses only on this hegemonic logic.

Neoliberalism is often analyzed materially in terms of the effects of expanding global capitalism. The approach adopted in this project does not deny these effects but focuses analysis primarily on the logics and technologies of government that have engendered them. In particular, this project emphasizes how risk has been figured by neoliberal logics and technologies.

American neoliberal logics of government valorize the market as the model for all realms of society. Foucault writes in *The Birth of Biopolitics:*

> The generalization of the economic form of the market beyond monetary exchanges functions in American neo-liberalism as a principle of intelligibility and a principle of decipherment of social relationships and individual behavior. This means that analysis in terms of the market economy or, in other words, of supply and demand, can function as a schema which is applicable to non-economic domains. And, thanks to this analytical schema or grid of intelligibility, it will be possible to reveal in non-economic processes, relations, and behavior a number of intelligible relations which otherwise would not have appeared as such—a sort of economic analysis of the non-economic. (243)

In order to illustrate how economic principles are applied to (formerly) noneconomic realms of life, Foucault describes how neoliberal thinkers characterized the mother-child relationship as an "investment" in "human capital" (243–44). The mothers' quality of care, affection, time, and pedagogical assistance for her child are all reduced to an economic formulation, an investment in human capital.

Foucault argues that in addition to extending neoliberal logics to spaces of life formerly regarded as private (e.g., the mother-child relationship), neoliberal authorities often use the market as "an instrument, a tool of discrimination" (247). The market becomes a tool in a curious kind of inversion that occurs. Under laissez-faire government, the market was a principle of government's self-limitation. But under neoliberalism, the market becomes a tool to be turned against government: "It is a sort of permanent economic tribunal confronting government" (247). These two aspects, the "analysis of non-economic behavior through a grid of economic intelligibility, and the criticism and appraisal of the action of public authorities in market terms" (248) are important dimensions of neoliberal governmentalities.

Neoliberals view competition as the factor driving market efficiencies. Donzelot explains in a 2008 essay that competition, or the "equality of inequality," is viewed by neoliberals as the mechanism that guarantees that markets allocate resources most efficiently (Donzelot "Michel" 123). However, unlike laissez-faire liberalism, neoliberalism acknowledges that competition is not in itself a natural phenomenon. Neoliberals thus believe the state should foster and expand competition among market actors to facilitate optimal market operations: "The role of the state is to intervene in favor of the market rather than because of the market, in such a way that the market is always maintained and that the principle of equality of inequality produces its effect" (124). Exclusions from the market must be eliminated, because optimal competition requires optimal participation. However, neoliberals believe that the state should never itself compete with the market. Thus, all forms of state-supported assistance should work on behalf of the market rather than in competition with it. Moreover, state expenditures aimed at enhancing market competition must be rationalized using cost-benefit analyses.

Given the constant market scrutiny of public authorities, the decision to preserve social welfare apparatuses requires elaborate rationalization using statistically calculable representations of increased efficiency and measurable outcomes. Accordingly, O'Malley and Hutchinson describe neoliberal shaping of risk management as characterized by "financial accountability" and "cost-benefit analyses" (374). Social welfare apparatuses remain viable within neoliberal logics only to the extent that they can deliver cost savings. For instance, the administration of justice has moved toward "loss prevention rather than moral rectification" in managing "social" problems (Reichman 151–72).

Neoliberal logics informing efforts to model risk remain actuarial but disperse responsibility for managing risks to individuals and privatized communities. O'Malley observes that although neoliberal logics share laissez-faire adulation of market autonomy, the former diverge in their valorization of particular discourses and practices organized around key signifiers such as "empowerment," "decentralization," "performance," "participation," "communities," and "prevention" (63). In effect, neoliberalism encourages individuals to assume responsibility for their employment, their health, their education, their life-long learning, their financial independence, and their happiness. Yet, as Mitchell Dean observes, managing risks is challenging for neoliberal subjects because of the politicization of the identification and management of risks:

> Individuals are left in the fateful situation of attempting to control their own proneness to risk by the use of expert knowledge that has been deprived of intrinsic authority. Indeed, "risk society" is something of a misnomer. Not only has expertise lost its intrinsic authority, but the kind of risks that we face

today are ones that often cannot be limited, are unpredictable, incapable of compensation and finally incalculable. (Dean *Governing* 65–66)

For example, parents seeking to weigh the relative risks versus benefits of using plastic baby bottles, as opposed to glass ones, to feed their children are confronted with expert conflict and indecision about the health risks and long-term health effects of commonly used plastics. Faced with the possibility, albeit remote, of irreparable damage to their infants, parents may go to extreme measures to find less risky alternatives even when those alternatives also pose potential risks, such as those posed by glass bottles or by pollutants in breast milk.

Neoliberalism—as a body of knowledge, strategies, and practices of governance—seeks to divest the state of paternalistic responsibility by shifting social, political, and economic "responsibility" to privatized institutions and economically rationalized "self-governing" individuals. By stressing individual "self-care," the neoliberal state relinquishes paternalistic responsibility for its subjects but, simultaneously, holds its subjects responsible for self-governance (Lemke 201). Neoliberalism encourages technologies of the self that govern from a distance, permitting "individuals to effect by their own means or with the help of others a certain number of operations on their own bodies and souls" (Foucault *Technologies* 18).

American neoliberal logics encourage societal assistance if that assistance is philanthropic in nature and is aimed at encouraging individual responsibility and market initiative. Philanthropy thus serves to extend the logics and capabilities of the market. This formulation of philanthropy's role in extending beneficent market operations implies that poverty stems from a failure of market penetration. Within neoliberal logics, poverty derives from a lack of market disciplines and market opportunities. In other words, poverty derives from barriers to market penetration of psyches, lifestyles, communities, and so on.

Neoliberal values and problem-solution frames encouraged the privatization of services and deregulation/liberalization of business, trade, and financial markets during the last three decades of the twentieth century. According to Mark Thoma, two fundamental neoliberal premises, the rational expectations hypothesis (REH) and the efficient market hypothesis, rationalized government deregulation and privatization. Thoma explains that REH derived from the work of two University of Chicago economists, Robert Lucas and Thomas Sargent, who argued in the 1970s that a market economy operates akin to a mechanical system governed by clearly defined, immutable economic laws. He adds that REH merged with a related theory of "efficient financial markets" that presumes that markets are populated by rational actors who set prices that reflect all available knowledge accurately. These theories discredited the role of regulators because the theories valorize market actors' knowledge and decision-making capacities over regulators'. In effect, these theories extended and rationalized applications of the classical liberal belief that through "his" pursuit of self-interest, homo oeconomicus guarantees the collective good, despite the invisibility of that good from the point of view of individual actors (see Foucault *Biopolitics* 282–92).

Deregulation of financial operations and institutions contributed to speculative bubbles in the first part of the twenty-first century, including the 2001 dotcom and 2007 housing crashes. These crashes led some critics to call into question the wisdom of complex derivatives, but they failed to produce greater public scrutiny of, or dialogue about, other neoliberal regulatory reforms, particularly those pertaining to labor, offshoring of production and finance, and the "great risk shift," a phrasecoined by Jacob Hacker to describe societal rejection of social welfare programs and technologies of government in favor of those that individualize risks and responsibilities (e.g., college education, job production and training, health, retirement, etc.). In the context of these developments, the position put forward in this project is that neoliberal logics of government continue to prevail across the realms of everyday life—of "civil society"—despite the election of a charismatic Democrat to the U.S. presidency.

I argue in the last chapter of this book that U.S. government policies to stabilize the economy are in fact driven by an overwhelming desire to restore the fantasy of the self-regulating market. Recognizing that government "interference" in markets runs counter to laissez-faire and neoliberal logics, the U.S. policy response has been primarily dedicated to backstopping private losses. Thus, the amount of money spent on bailing out financial and insurance institutions—$5,856,800,000,000—far exceeds stimulus funding aimed at jumpstarting business and revitalizing education—the $787,000,000,000 American Recovery and Reinvestment Act (Lavin and Bachman 58–59). In essence, the U.S. policy response has aimed to minimize the government's decision-making role while maximizing market forces. This response implicitly validates extant economic logics. Homo oeconomicus largely escapes blame for the crisis, because government interference with interest rates, fraud, and other barriers to transparency are cited by neoliberal media and pundit accounts as the causes of the collapse. Fraud and lack of transparency have emerged as the problem-solution frames that will dictate any regulatory reforms. This persistent belief in the efficacy of laissez-faire liberalism and neoliberalism will surely produce significant crises for the vast majority of children in America over the next twenty years.

PROJECT ORGANIZATION

Chapter 2: A Genealogy of Family Life and Childhood Governance. This chapter contextualizes the problem spaces of childhood and children within Western liberal logics of government (i.e., governmentalities), such as laissez-faire liberalism, social welfare liberalism, and neoliberalism, using Foucauldian genealogies of childhood, including works by Philippe Ariès, Jacques Donzelot, and Nikolas Rose; my own work; and Foucauldian and sociological analyses of risk. The chapter emphasizes how the child's soul and salvation emerged as central concerns of government across the nineteenth and early twentieth centuries and explains the economic and cultural factors contributing to public and philanthropic efforts to discipline and cultivate children's souls. Of particular interest are the governmental strategies used to identify, discipline, and "save" dangerous children during the nineteenth century and children constituted as "at risk" in the early twentieth century. Analysis of these strategies draws upon the excellent scholarship on childhood and extends existing analyses in the context of Foucauldian biopolitics. In sum, this chapter contextualizes the more focused analyses offered in subsequent chapters.

Chapter 3: Risk, Biopolitics, and Bioeconomics. This chapter explores how U.S. neoliberal economic and political logics that shift risk to individuals inform parenting, childhood education, and expert and everyday cultural attitudes toward childhood. The chapter explores the implications of neoliberal logics and practices in terms of an increasingly competitive global economy that promises upwardly mobile children "knowledge intensive" jobs while offering lower-income children few opportunities. The chapter contrasts the technologies of optimization used to govern affluent children with the technologies of normalization, seclusion, discipline, and punishment used to govern low-income children in the United States.

Chapter 4: Biopolitical Sorting Strategies. This chapter analyzes how risks posed to and by children were framed biologically across the twentieth century and into the twenty-first century. The chapter uses lead poisoning and geneticization of the diagnostic disorder ADHD (Attention Deficit Hyperactivity Disorder) to illustrate evolving approaches toward understanding and managing children's biological risks. The history of efforts to document lead poisoning demonstrates how epidemiological approaches to representing and managing risk were employed by medical authorities to promote social welfare projects. In contrast, contemporary medical and scientific representations of genetic risk tend to promote neoliberal frameworks for understanding and managing biological risks. Genetic approaches framing children's risks for disorders such as ADHD promote market logics by capitalizing upon medical-scientific "discoveries" and by encouraging individualized consumption of technologies of risk management by parents.

Chapter 5: Biopower, Security, and Development. This chapter addresses how U.S. security discourses and risk management technologies seek to govern poor children abroad and "illegal immigrant" children domestically. The chapter addresses the merging of security and development in totalizing governmental frameworks employed by the United States when evaluating the "demographic risks" of foreign childhood populations. The chapter examines how media representations of foreign children draw upon, and promote, formulations of childhood innocence and dangers that encourage foreign interference by the United States. The chapter concludes by examining the growth of disciplinary incarceration strategies used to govern the flows of immigrants populations into the United States prompted by neoliberal economic policies and reforms.

Chapter 6: Children and the Twenty-First Century. This chapter concludes the book by anticipating how the discourses and technologies of government discussed in this book might play out in the context of the current financial crisis. The U.S. financial crisis that began in 2007 may have profound implications for the government of U.S. children. With state and federal budgets being slashed and even greater competition for fewer economic and social opportunities being promoted, this crisis imperils pastoral biopolitics. The chapter also addresses the possibilities for a new economic order challenging the U.S. neoliberal one and discusses the implications of a new economic order for the government of populations domestically and internationally.

In sum, *Governing Childhood into the 21st Century* synthesizes and extends the disparate strands of scholarship on childhood, critical educational theory, biopolitics, and risk/governmentality studies while grounding them in the social contexts of everyday life. The project is designed for readers interested in a rigorous, comprehensive introduction to the wide array of interdisciplinary work focusing on the sociology of childhood and education, cultural studies of everyday life, Foucauldian biopolitics, and risk/security studies.

CHAPTER 2

A Genealogy of Family Life and Childhood Governance

The contemporary child is at risk from countless unseen dangers, while burdened with the responsibility for assuming the skills and knowledge necessary to achieve economic and social success. It is no wonder that childhood is today fraught with perils and risks, requiring expert guidance. Have children always borne so much risk and responsibility? In the naturalized context of contemporary life, it is difficult to appreciate the profound transformations in cultural practices and understandings that have produced Western childhood as a problem space requiring careful surveillance, deliberate cultivation, and expert guidance.

Philippe Ariès's *Centuries of Childhood: A Social History of Family Life* describes these transformations, identifying the new disciplines, social taboos, and surveillance strategies implicated in the production of modern childhood as a vulnerable social-symbolic space. Ariès, in accord with Jacques Donzelot and Michel Foucault, emphasizes the role of the Protestant Reformation in shifting attention toward the cultivation of the child's soul and the role of political economy in rendering the family visible as a source of national wealth through its economic productivity and reproductive capacities.

Ariès, Donzelot, and Foucault all see the Protestant Reformation as producing a cultural concern with children's souls. This cultural concern was historically novel and led to new child-rearing practices that essentially forged childhood as a unique space of life requiring special attention and special discipline. Western cultural concerns with, and practices of surveillance over, children's souls were reinvented across place and time in relation to children's salvation, their character, their future delinquency, their sanity, and their citizenship. Each articulation of the child, and the route toward her salvation, helped delineate norms of personhood, ideals of citizenship, and expectations for economic productivity in place and time. Each articulation of the child implicated idealized, normalized, and denigrated family contexts.

Foucault and Donzelot also both emphasize how children's economic importance within mercantile and liberal economic formulations encouraged new strategies for monitoring and governing child populations. Under state mercantilism, financial authorities began to link estimates of national wealth to the productive capacities of the population. This type of linkage encouraged greater surveillance over, and government of, the health of the population, particularly the health of children, who assumed importance in terms of their future productive capacities. Social reformers expressed great concern over the moral industriousness of family members and seized upon childhood as the proper problem-solution frame for reforming the unworthy poor by instilling discipline and character in children. Thus, the moral economy of childhood government was linked to the economic productivity of Western liberal nations.

Efforts to govern children's souls and future economic productivity have evolved in accordance with shifting historical circumstances. Genealogical analysis respects the complexity and contingency of local events but also seeks to discern regularities in historical conduct that might point to the circulation of common norms and technologies of government. This chapter addresses how distinct representations of childhood and technologies of government emerged in the United States in the nineteenth and twentieth centuries in response to the circulation of new valuations of their innocence and market utility. This chapter is particularly concerned with the creation, institutionalization, and normalization of "liberal" technologies for representing and managing risks to childhood innocence and economic productivity. What follows unpacks shifting technologies and strategies for representing and governing childhood, focusing on historical ontologies of childhood and practical pedagogies of childhood government.

The Invention of the Child

Ariès's analysis of French artifacts and historical records suggests that the idea of childhood did not exist in medieval society and that contemporary views of childhood began to appear more frequently in the seventeenth century. Ariès sees children's appearance "as children" signified by changes in their appearance within art. For instance, in the seventeenth century, children "of quality" were no longer dressed as adults (50). Moreover, Ariès argues that around the seventeenth century it became acceptable to speak openly of the unique pleasures of childhood as "coddling" made its formal appearance in the family circle (132). However, according to Ariès, coddling would soon disappear from historical accounts as clergy and pedagogues assumed new authority over childhood. These authorities aimed to produce disciplined, rational children and regarded them, at least in their writings, as "fragile creatures of god who needed to be both safeguarded and reformed," particularly in relation to their sexuality (133).

Children as Fragile Creatures of God

Transformations in Christian practices and understandings across the sixteenth and seventeenth centuries played a role in the emergence of this new conceptualization of children as fragile creatures of God. In particular, a growing Christian religious interest in the "inner" life of the soul would transform devotional practices, bringing children's spiritual life into focus as a critical problem-solution space. This concern over children's spiritual development would encourage greater scrutiny of children's daily practices and inner thoughts.

Jonathan Z. Smith argues that by the early eighteenth century, the meaning of Christian ecclesiastic vocabularies had shifted away from terms naming behaviors and groups, toward terms that referred to an interiorized and privatized inner life, requiring development of new strategies for evaluating the authenticity of religious experience (269–84). These strategies would embrace the soul and will, thereby prompting ecclesiastic surveillance over inward signs of devotion or transgression. The confessional, for instance, offered religious authorities greater access into pastorates' inner devotion while providing these authorities opportunities for acting upon transgressions through penance.

Foucault observes in *The History of Sexuality* that these new Christian vocabularies and technologies of government eventually centered on the family generally and children particularly. Foucault, like Smith, traces to these changes to the Counter-Reformation's efforts to impose ever more "meticulous rules of self-examination" and a more rigorous approach to penance upon the pastorate (19). Stricter and finer rules of self-examination had the effect of transforming the moment of transgression from the act itself to its formations in desire and will, particularly in relation to the formation of sexual desire. This concern about the formation of desire and the necessity of penance encouraged greater ecclesiastic involvement over the interior life of the pastorate and required that penitents transform their desire into discourse that could be analyzed and diagnosed. By transforming desire into discourse within a system of surveillance and confession, desire could be displaced, reoriented, and modified (23). Foucault claims this process produced a novel ecclesiastic interest in scrutinizing and redeeming souls. While adults were the initial target of this transformation, children eventually became subject to greater ecclesiastic interest, surveillance, and pedagogical practice.

Accordingly, Ariès notes that after the seventeenth century, sexual openness with children disappeared within Western European society in response to the moral reformation. Prior to this time, the idea of childhood sexual innocence simply did not exist (106). With the Reformation, new surveillance and disciplinary practices were implemented to protect childhood innocence. Childhood was to be safeguarded from "pollution by life," particularly by sexuality. Additionally, childhood was to be strengthened by "developing character and reason" (119). Thus, a "whole pedagogical literature for children" distinct from adult literature appeared in the seventeenth century to educate children in proper moral conduct. This literature encouraged confessional practices whereupon children were to express their misdeeds in order to repent. Moreover, Ariès observes that among the moralists and pedagogues of the seventeenth century, "fondness for childhood and its special nature no longer found expression in amusement and 'coddling' but in psychological interest and moral solicitude" (131).

In his psychoanalytic history of childhood, Lloyd DeMause describes the form of child rearing that grew out of the religious reformation and pedagogical orientation as the "intrusive mode" of child rearing. He describes this mode as shifting the focus of control from the ambivalent physical molding of children found across the fourteenth through sixteenth centuries to their psychic molding, beginning in the seventeenth century. The intrusive mode assumed childhood innocence but simultaneously introduced more invasive forms of child governance:

> The child was no longer so full of dangerous projections [of the devil], and rather than just examine its insides with an enema, the parents approached even closer and attempted to conquer its mind, in order to control its insides, its anger, its needs, its masturbation, its very will. (52)

This intrusive mode of child rearing was central to producing the moralistic, pious, economically productive subjects of early liberal capitalism. The earlier brutally repressive forces used to control populations had to be supplemented with new technologies that elicited and shaped the life forces of vulnerable and malleable populations.

The intrusive mode of child rearing can be seen in approaches to children's spiritual salvation in the colonial United States, although the Puritans were slower than Europeans to shed older ideas that children were depraved from birth (Wollons x). Puritans' fears that children risked eternal damnation if not properly guided toward salvation led to meticulous practices of child surveillance aimed at instilling absolute obedience, self-discipline, and self-denial (x). The difficulty of colonial life no doubt contributed to strict and often punitive treatment until new ideas about raising children began to circulate around the turn of the nineteenth century. Still, even after the circulation of new ideas that encouraged less punitive child governance, breaking children's wills remained an important objective of American evangelical child-rearing practices (Greven 86–87). Indeed, Greven observes that evangelicals sought to break children's wills very early, from infancy onward, to "make the child's obedience habitual and 'natural'" (93). Evangelical discipline was most effective when it was internalized and exerted by the child herself as the "inescapable inner disciplinarian" (93).

Around the turn of the nineteenth century, improvements in standards of living and the spread of liberal ideals contributed to a degree of relaxation of strict and often punitive attitudes toward child rearing in the United States, encouraging people to see children as fragile creatures requiring gentler cultivation. The private family—constituted by father, mother, and children—had emerged by this time, allowing for tighter emotional bonds among family members and more deliberate cultivation of children's character (Mintz 49). Efforts to shepherd children's character through careful government were not limited to the home, as schools emerged as sites for the cultivation of childhood character. For example, in a 1798 text, Benjamin Rush (1745–1813) argues for reforms in student education and discipline. Rush writes that students "fhould be permitted, after they have faid their leffons, to amufe themfelves in the open air, in fome of the ufeful and agreeable exercifes which have been mentioned. Their minds will be ftrengthened, as well as their bodies relieved by them" (62). Rush urges that the reduction of "cruel" corporal punishment by civil, ecclesiastic, and military authorities should be extended into the schools. Rush contends that the "fchoolmafter remains the only defpot now known in free countries" (63). He suggests instead a gentler demeanor entailing schoolmasters' efforts to cultivate respect and affection. "Private admonition" and after-school containment should replace beatings (69). In sum, Rush advocated for a more benevolent pastoralism for educational governance focusing on molding the child's inner life, particularly his or her conscience.

Ariès observes that this shift in child rearing was delineated by a further reduction of corporal punishment: the "insistence on humiliating childhood, to market it out and improve it, diminished in the course of the eighteenth century" (262). Children were to be cultivated, not simply molded through physical force. Moreover, the idea that children required cultivation to overcome original sin slowly ceded to the idea that cultivation shaped malleability and innocence (M. Grossberg 8).

John Locke (1623–1704) and Jean Jacques Rousseau (1712–1778) are historically significant for promoting pastoral government of childhood innocence. Locke observes in *Some Thoughts Concerning Education* that child rearing often fails to make the "Mind" "obedient to Discipline and pliant to Reason" (75). Locke represents the development of reason in childhood as necessary for the self-government of "Appetites" in adulthood. Rousseau's texts emphasize the role of pedagogy in developing the capacities of self-government. Accordingly, he advises the tutor: "Let your pupil always believe that he is the master, but in fact be the master yourself. No other subjection is so complete as that which keeps up the pretense of freedom; in such a way one can even imprison the will" (cited in E. Singer 39). Efforts to cultivate an "active conscience" and the capacities of self-government thus became important goals of parenthood and childhood education (Greven 93). Still, adoption of more pastoral and less corporal means of discipline varied widely by place; for instance, lessening of punitive discipline did not occur in England until the nineteenth century, nearly a century after reforms in France. Moreover, romantic ideals about children did not appear frequently in American children's literature until the mid-nineteenth century (Macleod 97), and American conservative evangelicals encourage corporal disciplining of children today.

The gradual, albeit uneven, emergence of children as "fragile creatures of God" (Ariès 40) in religious doctrines and artifacts roughly coincided with new concerns about the family driven by mercantile formulations of wealth and nationhood. The family emerged as an important focus of government in European societies by the eighteenth century because it was linked to national wealth and because it was regarded as the central site for the production of citizens. The health of the family and its productivity, therefore, became problem-solution frames that concerned both state and early philanthropic authorities in European societies, especially over the course of the nineteenth century.

Children, Political Economy, and the Birth of Liberalism

New mercantile formulations of wealth that derived value from the productive capacities of populations encouraged new approaches toward childhood governance. Tracts began to appear aimed at educating the (literate) populace on practices capable of reducing childhood mortality. For example, a 1733 tract, *The Art of Nursing: Or the Method of Bringing up Young Children According to the Rules of Physick For the Preservation of Health, and Prolonging Life* promises to "oblige the Publick" by explaining "The true Way of bringing up young Children" (Brotherton and Gilliver 3). William Moss published "An Essay on the Management, Nursing and Diseases of Children, from the Birth: And on The treatment and Diseases of Pregnant and Lying-In Women, which was "defigned for Domeftic Ufe, and purpofely adapted for Female Comprehension" in 1794 (title page). Issues pertaining directly to the growth of the population—including midwifery, nursing, treatment of childhood diseases, and the practices of foundling hospitals—were of central importance in these texts.

Population growth mattered to mercantile authorities because it enhanced the nation's productive capacity. Mercantile concern over the balance of trade made enhancing productive capacity a central objective of state government, producing this new interest in the population's fecundity and health. Therefore, new medical and public health authorities were created and empowered with the authority to "police" the health and economic productivity of the population under state mercantilism (Foucault *Biopolitics* 7). Public sanitation was enhanced, orphanages founded, and medical hospitals instituted to promote the fecundity and productivity of the nation. However, the internal police power of the state eventually met limits articulated in the doctrines and practices of political economy, which began to circulate toward the end of the eighteenth century.

In the eighteenth century, political economists such as Adam Smith (1723–1790) advocated for limits on sovereign authority. Drawing upon John Locke's premise outlined in *Second Treatise of Government* that individuals own their own bodies, political economists suggested that market transactions involving personal property should be relatively free from sovereign interference. In this way, political economists carved out the "market" as a social space outside the purview of sovereign authority. The market was not regarded as an anarchic space, because its characteristic transactions were believed to governed by natural laws. Liberals held that individuals should be "free" to operate autonomously in the market because these natural laws guaranteed the orderly character of market transactions and promised their mutual beneficence. The market therefore emerged as a private space ideally free of state control, while liberal ideology elevated the principle of market exchange as the ultimate expression of personal and societal freedom (Foucault *Biopolitics* 47).

At this time, the idea of civil society also emerged within liberal philosophy. This privatized realm was invented as a semi-autonomous space wherein the population could exercise personal freedoms of expression (see Dean *Constitution* 13). The emerging nuclear "family" was an important central feature of civil society. Within the confines of the family structure and household, the population was encouraged to exercise freedoms tempered by morality and prudence. Under late eighteenth-century and nineteenth-century liberalism, the virtuous, self-determined, industrious, forward-thinking individual was the idealized inhabitant of this semiprivate sphere of family/population, bridging micro and macro cosmos. However, the relationships, institutions, and apparatuses of civil society and the market fundamentally limited the freedoms they produced. As Foucault explains: "Liberalism must produce freedom, but this very act entails the establishment of limitations, controls, forms of coercion, and obligations relying on threats, etcetera" (*Biopolitics* 64). Liberal freedoms are therefore always/ already characterized by liberal limits and constraints.

Western states' mercantile interests in policing the population and controlling commerce at times conflicted with liberally defined market freedoms.[1] In particular, Foucault describes how the naturalization and universalization of free-market exchange led economic liberals to contest the utility of the state's police apparatuses. Thus, the "question of utility" operated to limit pastoral state involvement with the populace (*Biopolitics* 47). For example, in the context of nineteenth-century state-directed efforts to improve public sanitation, liberals pressed for accounts of the utility of state expenditures. What economic value derives from sanitation efforts? What value accrues from vaccination? What value derives from the formation and maintenance of orphanages for children? Liberal political economy demanded that governmental (i.e., police) operations aimed at enhancing the biovitality of the population be rationalized within economic calculi of value. Moreover, liberal political economy demanded that police operations be limited in scope to that social space defined as "private" to preserve the essential autonomy of the market. The welfare of the populace is less important in liberal formulations than the freedom of the market.

Liberals believed that private philanthropy promoted the biovitality of the population with fewer infringements on individual and market autonomy than state operations. Private philanthropy operated without the infringement of public taxes and state intrusion and therefore arose within the liberal imagination as the most desirable technology for shepherding families and communities who failed to meet liberal standards of economic sufficiency until the mid-twentieth century. Additionally, the religious character of philanthropic enterprises promised the inculcation of moral values and thereby assuaged concerns that aid would bequeath sloth.

Public and philanthropic efforts to cultivate the population's economic vitality and personal morality converged in a nineteenth-century security complex aimed at children's welfare and education. In *The Policing of Families,* Donzelot describes this security complex as "tutelary" and argues that its main objective was pacification of the population. Pacification strategies often involved separating children from their parents. Accordingly, Donzelot describes how passage of a series of bills in the mid-1800s aimed at protecting children from economic exploitation or abandonment by their itinerant parents led to the formation of specialized institutions responsible for their moral transformation into good citizens of the state. Children were thus regarded as the crux of social reform, as illustrated by this passage from an authority of the time: "So long as society will not begin this reform at the base, that is, through an untiring vigilance over childhood education, our manufacturing cities will be constant centers of immorality, disorder, and sedition" (cited in Donzelot 72). Concern over child welfare did not reflect a concern over the quality of life for the child per se, but rather reflected the view that careful control over the child-adult relationship was vital for protecting cities from working-class revolt. Relief was therefore pedagogical in intent; it aimed at disseminating desirable social norms of industrious morality (73). Thus, Foucault summarizes in *Psychiatric Power* that assistance was designed to reinscribe and supplement social structures believed to promote order and obedience:

> In short, the function of everything we call social assistance, all the social work which appears at the start of the nineteenth century, and which will acquire the importance we know it to have, is to constitute a kind of disciplinary tissue which will be able to stand in for the family, to both reconstitute the family and enable one to do without it. (*PP* 84)

In order to bolster or substitute paternalistic family structures, the tutelary complex drew upon and reinscribed older philanthropic and hygienic institutions.

The nineteenth-century tutelary complex was acceptable to liberal sensibilities because it promised utility through the inculcation of a prudent, industrious morality among potentially dangerous children. The public and private authorities who comprised the tutelary complex circulated widely, and their protocols and technologies for child government were eventually adopted by families pursuing liberal ideals of personhood. The tutelary complex discussed in this chapter evolved over time and place, but its characteristic focus has been the reduction of biopolitical risks posed to and from children. Efforts to govern and reduce the risks associated with children can be regarded as strategies of security.

As Foucault explains, "strategies of security" are the other side of liberal freedoms (*Biopolitics* 65). State-supported and philanthropic security apparatuses promised to govern the risks to freedom and to arbitrate between individual and collective interests. But the security apparatuses stimulated perceptions of danger and risk by identifying threats posed to and by individual liberties (66). Thus, Foucault observes that an entire "education and culture of danger appears in the nineteenth century" (66). The culture of danger manifested in journalistic interest in crime, campaigns around disease and hygiene, and fears of degeneration. These accounts stressed the ubiquity of threats, thereby replacing centered, apocalyptic views of danger that had existed previously with diffuse and circulatory accounts. Foucault concludes, "There is no liberalism without a culture of danger" (67).

The culture of danger and the preoccupation with security promoted extension of "procedures of control, constraint and coercion," particularly for children who emerged in the nineteenth century as essential risky subjects (*Biopolitics* 67). In a seeming paradox, the nineteenth-century age of liberal freedoms was accompanied by "development, dramatic rise, and dissemination throughout society of these famous disciplinary techniques for taking charge of the behavior of individuals day by day" (67). Children emerged as important subjects of new mechanisms aiming to produce and securitize freedom through additional control and intervention. Children's discovered vulnerability encouraged creation of specialized institutions (orphanages, juvenile courts, clinics) and pedagogical authorities and technologies, which operated together to monitor, cultivate, discipline, and socialize their development. The next section explores these efforts to govern children through the development of a "tutelary complex" (Donzelot *Policing* 9).

Governing Children in the Nineteenth Century: Constitution of the Tutelary Complex

As explained above, liberalism both produces and limits freedoms. Liberalism limits freedoms by encouraging the establishment of procedures of control, constraint, and coercion. The mandate of security coupled with the ubiquity of danger warrants apparatuses of control. Apparatuses of control do not simply operate coercively but also operate productively by eliciting and channeling the population's life forces. As mentioned in the previous section, children became critical sites of danger and targets of control in the nineteenth century. Their newly discovered innocence, vulnerability, and malleability mandated careful parental governance; their potential degradation by exposure to contaminating influences necessitated philanthropic and state interventions.

The tutelary complex that arose to govern children incorporated the expertise and resource of an older philanthropic relief/assistance complex as well as the knowledge and technologies of nineteenth-century medical-hygienist authorities. A wide array of medical, hygienic, and philanthropic authorities were therefore mobilized to assist in the child's socialization. Their involvement in childhood government would lead to the "discovery" of norms of health, norms of conduct, and norms of morality believed to be vital for the child's proper socialization (Donzelot *Policing* 57). Foucault sees the

"Psy-function" comprised of psychological institutions and authorities as playing a particularly important role in creating disciplinary apparatuses (by way of normalizing knowledges and practices) believed capable of patching the growing ruptures in familial sovereignty (*Psychiatric* 85; see also Rose *Governing* 1–272).

DeMause therefore describes the articulation and promotion of norms of childhood government as producing a new mode of childhood—the "socialization mode"—which he claims ranged from the nineteenth century through the mid-twentieth century (52). Like Donzelot, DeMause identifies this historical period as emphasizing childhood *normalization*. Although the "socialization mode" maintains an interest in children's souls, it is characterized primarily by the emergence of "expert" knowledge about child rearing and the dissemination of formalized regimes of child discipline and character development. The socialization mode and the corresponding tutelary complex became evident in new attitudes toward mothering and in new institutional forms aimed at rehabilitating dependent children.

The Socialization Mode and Responsible Mothers: Circulation of Security

Nineteenth-century anxieties about overpopulation of the dangerous classes and new technologies of childhood socialization together transformed responsible motherhood. Ron Greene chronicles how Thomas Malthus resolved his pessimistic problematic of population growth outstripping agricultural output with the responsible couple who would voluntarily restrict their reproduction through "'preventive checks' such as delayed marriage, voluntary celibacy, and abstinence in marriage" (Greene 20). As Greene explains, Malthus supported educating working-class parents for moral restraint to securitize the nation against their overpopulation while arguing against any form of state intervention in the economy that would disrupt laissez-faire market operations. The family's sexuality would ultimately serve as a focal point for surveillance over, and securitization of, the nation's health and its reproductive suitability in the latter half of the nineteenth century. The onus and responsibilization of procreation centered on the privatized couple, in particular, the mother (104–105). The mother would assume responsibility for controlling her procreation and for governing her children's socialization through careful character education focusing on prudence and a strong work ethic. Families who failed to fulfill these responsibilities were subject to greater philanthropic and public scrutiny.

Steven Mintz argues that mothers' role in socializing children in the United States can be traced to the immediate post-Revolutionary war period when moralists and ministers invested mothers with the responsibility for their children's fulfillment of republican virtues (71). Bourgeois women in the United States therefore came to understand their class status in relation to their moral restraint in child bearing and their cultivation of quality children. Women in America were able to restrict their reproduction through contraception legally until the "purity" campaigners of the 1870s led drives for its outlaw (M. Grossberg 175–76). By limiting their reproduction, women could provide more quality care. This emphasis on quality helped reconceptualize the economic value of children as developable assets, mobilizing efforts by mothers to seek out expert advice on the care and proper socialization of their children (see also Zelizer 1–228).

For instance, John Abbott's "The Mother at Home," originally published in 1834, advises mothers that they hold primary responsibility for formation of a "virtuous character" in their children (84). Mothers in this text are interpellated, or hailed, as both "guardians and controllers of the human family" (84).[2] Evidence of mothers' willingness to assume responsibility for their children's virtue is illustrated in this account of mothering written by an American woman in 1813:

> There is scarcely any subject concerning which I feel more anxiety, than the proper education of my children....The person who undertakes to form the infant mind, to cut off the distorted shoots, and direct and fashion those which may, in due time, become fruitful and lovely branches, ought to possess a deep and accurate knowledge of human nature. It is no easy task to ascertain, not only the principles and habits of thinking, but also the causes which produce them. It is no easy task, not only to watch over actions, but also to become acquainted with the motives which prompted them. It is no easy task, not only to produce correct associations, but to remove improper ones, which may, through the medium of those nameless occurrences to which children are continually exposed, have found a place in the mind. But such is the task of every mother who superintends the education of her children.(Cited in Reef *Childhood* 16)

These statements illustrate how the project of nineteenth-century child rearing involved careful monitoring of the sources of action and deliberate interventions designed to elicit desired outcomes. New security technologies were born and circulated as child rearing became the organizing focus of societal normalization. Mothers played a vital role in adopting and executing these new security technologies.

One primary way that expert authority informed mothers' child rearing was through medical expertise. A growing body of medical expertise informed mothers how to feed and care for their infants. Sanitary practices would play a particularly significant role in shaping nineteenth-century bourgeois motherhood. By the mid-nineteenth century, women were instructed by sanitary authorities in household sanitary practices designed to create boundaries between contaminated external spaces and internal pure spaces (Tomes 1–368). By the close of the nineteenth century, newly created hygienic norms provided specific corporal disciplines that mothers could follow to ensure their offspring's protection against germs and other social contaminants. The importance of these developments for shaping new attitudes about child rearing will be discussed later in this chapter.

In sum, across the nineteenth century, child rearing documents and women's magazines interpellated bourgeois women as responsible for transforming their children into virtuous citizens and "competent burghers" (M. Grossberg 8). This focus on the importance of respectable families in producing worthy citizens also raised the specter of unworthy families who threatened national well-being by their idleness, intemperance, and delinquent children (10). Divorce, desertion, male licentiousness, and women's rights were cited as behaviors that threatened national well-being (10). These concerns would bring the children of the poor into greater focus both in Europe and the United States.

Governing Dangerous Children: Strategies of Containment

The poor have long been regarded as a problem of government, but prior to the eighteenth century, poor children were rarely singled out as posing distinct problems. In the sixteenth and seventeenth centuries, European societies often resolved the problems posed by the poor through systems of "outdoor relief" that provided minimal subsistence to poor populations. Children living with impoverished parents survived on their parents' temporary work, occasional outdoor relief, and the meager wages of their own labor. Orphanages and hospitals that received children existed in some European localities but were not widespread institutions until at least the eighteenth century. But by the nineteenth century, the children of the poor became a central societal concern in Europe and the United States.

In early nineteenth-century America, household industries that had formerly employed children as laborers were increasingly replaced by factory enterprise, and the apprentice system was in decline (Mintz 155). These changes converged with population growth to increase the circulation of indigent children. In the United States, as in France, a tutelary complex comprising public and private authorities and institutions emerged to govern these children. This complex aimed both to protect children from society and to protect society from dangerous children (Mintz 155).

Concerned about destitute child paupers, early nineteenth-century American almshouses and workhouses (established as substitutions for outdoor relief) began to allow children entry. Although these institutions tried to rehabilitate adult intemperance and indolence, they lacked any form of pedagogical devices for rehabilitating children. While this deficiency would not have presented much concern in earlier years, new sensibilities about childhood vulnerability and malleability caused concerned activists to institutionalize specialized orphan asylums, which promised rehabilitation for child paupers.

Matthew Crinson's *Building the Invisible Orphanage* chronicles the evolution of specialized orphanages for children across nineteenth-century America. He argues that early nineteenth-century philanthropic Americans responded to perceptions of an increase in indigent poor populations, erosion of childhood discipline, and juvenile delinquency by attempting to replace outdoor relief with orphan asylums (37–39). The goal of the early orphan asylums was to remove children from corrupting social influences such as poor, intemperate parents and the "corruption of city life" (68). Many of the Protestant orphanages aimed to Americanize children of poor immigrant parents. Crinson describes the orphanages as highly regimented and often brutally disciplined.

Judith Sealander points out that many of the children in orphanages in the United States were not truly orphans; many parents were obliged to search for work and could not care for small children (98–112). Economic hardship afflicted many Americans in the nineteenth century, and working-class parents were often forced to migrate from town to town in search of work (see also Katz 7). Women, in particular, received such low wages that they could not hope to sustain themselves and their children on available work (Katz 7). Additionally, single mothers who found work as live-in servants were typically barred from bringing their offspring to work (Sealander 98–101). Destitute workers unable to receive sufficient support from kin or outdoor relief could be auctioned off as indentured servants. Poorhouses were typically a last resort, aimed primarily at keeping the poor from starving. Although some poorhouses emphasized order, routine, and self-sufficiency, many simply became overcrowded refuges for the very sick, the very old, and the mentally ill and thus became increasingly regarded as unacceptable locales for children who could potentially be rehabilitated. Destitute children were therefore diverted to orphanages if they could not be apprenticed or otherwise employed.

Beginning in the nineteenth century, children who committed crimes were also subjected to institutional segregation. Rather than being released to their parents' care, children caught in criminal activities began to be incarcerated in state penitentiaries. This type of incarceration worried some charitable and religious authorities, because children would be exposed to the worst forms of contaminating influences. Spurred by the Society for the Prevention of Pauperism, the state of New York approved construction of a House of Refuge for delinquent children in 1824, helping to pave the way for specialized institutionalization of juvenile rehabilitation. New York's House of Refuge may have been the first of its kind in the United States, but it represented a broader effort to create highly specialized institutions for children in the early nineteenth century.

Specialized asylums were also created for blind and variously disabled children, particularly after 1830 (Trent 12). During the first part of the nineteenth century, these specialized asylums tended to be rehabilitative and were inspired by the apparent successes of specialized European schools. Edouard Seguin's (1812–1880) apparent successes training "idiot" children inspired many similar efforts in the early nineteenth century. Seguin considered idiocy as a disorder of will and sought to educate "idiot children" using "moral treatment" (cited in Foucault *PP* 215). Foucault observes in *Psychiatric Power* that Seguin believed that the problem with "idiot" children was their "monarchical will," which entailed the child's refusal to submit to parental authority and to integrate within a system. Thus, the idiot was one who "stubbornly says 'no'" (215). Seguin saw the teacher's role as enforcing compliance by becoming the "absolute master of the child," and this mastery must be instantiated and enforced by the impeccable corporeality of the teacher's body, which is capable of subduing and disciplining the bodily energies of his students (216). Seguin's approach to tutelage required transforming the pedagogical social space into an institutional, disciplinary space resembling that found in the institutional asylum.

Seguin's special schools in France for "idiot" children inspired their institutionalization in the United States. Some focused more on "normalizing" "defective" children through specialized educational curriculum, while others simply trained their charges to engage in menial labor. Trent's *Inventing the Feeble Mind* chronicles the development and transformation of these schools in the United States and illustrates how they contributed to the creation of new classes of experts—educators, administrators, and psychiatrists (to a lesser extent)—who studied and governed these children. The function of these experts and their institutions shifted in the latter half of the nineteenth century, when childhood "idiocy" became increasingly associated with lack of morality, criminality, and degeneracy. For example, Trent argues that after the American Civil War, American conceptions of "feeblemindedness" diverged from the educational model promulgated by the French, as Americans increasingly regarded the "feebleminded" as burdensome, morally degenerate, and pathological. So began the transformation of the child "idiot" or "imbecile" into the "feebleminded," "delinquent" child, which entailed use of scientific medical models to explain the "degenerate" and "heritable" "pathologies" of the poor and afflicted (Trent 16). The "threat of contagion" from social undesirables altered the institutional goals of many late nineteenth-century asylums in the United States and Europe (Sibley 25).

The threat of contagion served as an important impetus for nineteenth-century reforms aimed at segregating and eventually rehabilitating children. However, by the close of the nineteenth century, *preventive* aspirations would supplement the goals of segregation and rehabilitation. As Stevenson explains, nineteenth-century philanthropy regarded "delinquency, dependency, and mental disease" as "interlocking fields" (9), but efforts to redress these ills underwent transformations, "from segregation to rehabilitation to prevention" (9). Accordingly, by the close of the nineteenth century, child savers pursued state-supported reforms not only to rehabilitate the children of the poor and act upon their social circumstances and habits, but also to prevent the development of *future* criminal and mental pathology. As Lyman Alden expresses in 1893,

> The aphorism, "Prevention is better than cure" is an old one; but it is only in recent times that the truth embodied in it has been practically applied.... And so, as never before, throughout all the civilized world the attention of philanthropists and social scientists, during the past forty years, has been turned to the prevention of social evils by caring for and properly training

that large class of destitute and neglected children from which the pauper and criminal classes largely spring. (68)

Governing Dangerous Children: The Circulation of Prevention

The idea that illness or crime is preventable—not simply subject to fate, God's will, or hereditary influences—was not widely accepted throughout much of the nineteenth century. Darwinian-inspired pessimism about inborn character defects and heritable degeneracy prevailed throughout much of the late nineteenth century. Still, a variety of individuals inspired by science or their faith did believe that preventive actions could reduce the chances of future criminality and save lives, souls, and public funds. Ideals and practices promoted by the sanitarians[3] provided a model of hygienic prevention that shaped nineteenth-century child saving, producing the conditions of possibility for early twentieth-century child guidance. This model would eventually inform the tutelary complex's approach to the government of (potentially) dangerous children. Child savers found it easy to appropriate the model because sanitarians had used child subjects when legitimizing the utility of their preventive ethos.

Sanitary reform began in England in the 1830s in response to the urban diseases and crowding brought upon by industrialization. Richard A. Meckel argues that sanitarians took a middle road between (1) social conservatives who viewed urban disease and poverty as stemming from moral degeneration caused by the "civilized" temptations offered by city life to simple migrants and (2) economic radicals who viewed urban disease as stemming from industrial exploitation (14–15). The sanitarians mediated these positions by offering an explanation of urban poverty and illness stressing environment (but not industry) and hygiene (15). Meckel explains that sanitarians believed

> filth caused disease and that disease promoted poverty and vice, they argued that social ills and the cost of relieving them could most effectively be reduced by sanitizing the urban domain through facilitating sewage and garbage removal, cleansing city streets, providing clean water supplies, and improving ventilation in both individual domiciles and the areas they occupied. (15)

The sanitarians' promotion of sanitary practices occurred well before bacteria were discovered as disease vectors.

Sanitarians used statistical data linking mortality to specific urban places to provide evidence for their arguments for sanitary reforms. Their data revealed higher mortality rates in urban areas populated by poor populations than in other areas (Meckel 15). The urban poor's potential as a source of contagion threatened the nation's health. The economic costs of their illness also figured in sanitarians' formulations because the ill poor could no longer contribute to national productivity and could appeal to public charity for their sustenance. Based on these economic arguments, the Sanitarians were able to generate broader support for their urban reforms that focused on improving the circulation of air and water and removing what were perceived to be contaminating influences. The sanitarians helped promote security technologies that operated preventively upon and through the population to promote health and productivity.

As mentioned previously in this chapter, infant mortality emerged as an important problem-solution space in the context of concerns about sanitary reforms and the urban poor (Meckel 19). Medical practitioners, motivated by the goals of political economy, had from the eighteenth century onward pondered strategies for preventing infant mortality in order to promote national productivity. High incidences of urban infant mortality were therefore documented by sanitarians in the mid-nineteenth century as one strategy for mobilizing public support for sanitary reforms (Meckel 5). From about 1850 to 1880, infant mortality was believed to be caused by hereditary debility and exacerbating morbific environmental influences; therefore, sanitarians pressed for general sanitation of urban environments (5). However, beginning in the 1860s, medical practice in Germany and England increasingly emphasized the importance of infant feeding in reducing child mortality (45). Accordingly, by the 1870s, American sanitary reformers stressed improper nutrition as playing a greater role than foul air and inherited debility (62).

Declining rates of breastfeeding and declining medical confidence in women's ability to successfully breastfeed made infant feeding a relevant topic for sanitary authorities. The newly professionalized specialization of pediatrics, established in the United States in the 1880s, was particularly important in shaping public and philanthropic officials' concerns about infant feeding practices (Meckel 52–53). Pediatricians debated optimal feeding practices and infant formulas, leading eventually to the creation of expert norms for the proper feeding of infants and to efforts to identify potential urban sites where dangerous feeding practices might occur. Sanitary reforms added impetus to, and fostered, pediatricians' child-saving programs such as infant-feeding stations.

Pediatric physicians adopted the newly developed germ theory of disease to spearhead their efforts toward social reform through the establishment of publicly funded "safe" milk supplies aimed at curtailing high rates of infant mortality attributed to unsanitary milk. This medical-hygienist strategy of intervention would engender subsequent tutelary interventions aimed at educating "ignorant mothers" about proper feeding at the stations, ameliorating the social contagion of the "adverse" home environments of immigrants, poor populations, and others deemed incapable of proper self-governance in the early 1900s, when strategies to prevent mortality expanded to encompass mothering behavior (Meckel 5; Nettleton 98–111). Accordingly, feeding stations ultimately adopted a broader approach toward educating mothers, one that was aimed not only at preventing infant mortality, but also at preventing future child delinquency. Home health visitors who entered immigrant women's homes to supervise their mothering also contributed to normalizing child rearing in relation to the advice of expert authorities.

The stated goals and practices of institutions for children evolved in accord with the preventive impulse. The late nineteenth-century juvenile asylum, for instance, gradually accepted a wider range of children, including those who had committed no offenses but whose degraded parents suggested future delinquency in the absence of tutelage (Crinson 69): "In general, the asylum was to reach and rescue 'that class, who were in danger of becoming, but who had not yet become, *Incipient Criminals*'" (69). Pauperism and vice were two sides of the same contagion (70). Tutelary efforts to combat criminality sought to inculcate "wholesome moral influences," desirable "habits," and strong moral "character" (cited on 52).

Although specialized institutions were designed to protect children from contaminating adult environments, beginning in the 1850s, some child activists began to question their effectiveness(Crinson 94). Charles Loring Brace (1826–1890) was particularly vehement about the dangers institutionalization posed to children's character development. Brace and other observers felt that institutional life encouraged surface adherence to rules and regulations but allowed secretive vices (Crinson 96). Institutional life was perceived by these reformers as machinelike, and they argued that it led to concealment of resentments and evil impulses (96). Reformers urged that institutional practices be altered to allow for greater individuality and the incorporation of family-like disciplines to replace those of the institutional machine (116).

Brace founded the Children's Aid Society in 1853 to assist poor children and to combat the newly discerned dangers that institutionalization posed to the child's character and capacities for self-governance (Brace 1–36). He felt that the disciplines of family life were superior to the machinelike disciplines of institutions (Crinson 116). He favored farm life, believing its disciplines were particularly well suited for fostering moral self-government (Finn 93–94). Therefore, orphan trains carried urban children to foster families in agricultural areas from the 1850s through the early 1920s. Families needing additional farm labor tended to be the major source of adoptions.

Preventive ideals prompted public and philanthropic efforts to remove children from undesirable homes before children were adversely impacted by their parents' failings. Nineteenth-century legal reforms in the United States allowed courts to dictate such removals using the idea of the child's best interest, or "nurture," as justification (M. Grossberg 237–39). One way that authorities gained access to private homes was through the "discovery" of child abuse. In 1874 there were no laws protecting children from abuse in New York, causing a city social worker to look to the American Society for the Prevention of Cruelty to Animals for assistance in rescuing a terribly abused nine-year-old girl. Harry Ferguson observes that by the 1890s, shelters had been created for abused children and that these shelters were "historically novel institutional spaces within which the body of the mute child was made available to welfare practitioners to be read off for signs of what was classifiable as child maltreatment in social practice" (35). Antivivisection activists, animal-welfare activists, and eventually the Federal Children's Bureau (founded in 1912) all sought to prevent the newly recognized conditions of child abuse and neglect in the United States.

Progressive child savers would eventually institute the first U.S. juvenile court in Cook County, Illinois, in 1899 using the British legal doctrine of *parens patriae*, or "the State as parent," which extended the state's authority over the welfare of vulnerable, at-risk populations. Michael Grossberg explains that the institutionalization of parens patriae made a father's custody rights contingent upon his responsibilities as a guardian rather than deriving absolutely from the father's property rights (236). This transformation signaled new interpretations of childhood innocence. Innocent children required new, legal protections in order to prevent victimization and exposure to contaminating influences.

Yet legal and philanthropic innovations and interventions aimed at preventing children's future delinquency and degeneration were not simply imposed upon families. Van Krieken argues that working-class parents often solicited help from the child-welfare officials of the tutelary complex in order to deal with "uncontrollable" and "stubborn" children (414). Moreover, Van Krieken contends that the "respectable" working classes were committed to the education and moral improvement of their children as means for social advancement (410). By disputing Donzelot's reading of the tutelary complex as exerting social control, Van Krieken illustrates how the liberal ethos of self-governance was disseminated across the social field as working-class parents sought respectability and mobility through the "improvement" of their children. Guiding both expert (social workers' and juvenile officials') and parental understandings of the proper means for moral and education improvement were new cadres of experts in the emerging professions of education, psychology, and psychiatry.

Medicalizing and Educating Dangerous Children:
Normalizing Deviance

As pediatricians professionalized and began sharing their accumulated experiences in specialized journals such as the *Archives of Pediatrics* (founded in 1884), children emerged within their medical gaze as a unique population, distinct from adults (Meckel 47). Unique maladies of childhood were "discovered" and documented in medical textbooks and the newly specialized journals. Perhaps more importantly, medical specialists also began to discern the origins or precursors of adult disorders, particularly psychiatric disorders, in childhood. This discovery that adult illnesses had precursors in early childhood contributed to the ideals of prevention that were fostered by reformers. Hygienic interventions and education thus assumed import for their roles in the prevention of adult mental "degeneracy" (i.e., mental illness) as well as criminality.

Prior to the 1880s, psychiatric interest in children was very restricted and primarily addressed masturbation in boys and precocity in girls as precursors to adult insanity (Jones 16). However, in 1879, Henry Maudsley (1835–1918) titled a chapter in one of his books "The Insanity of Early Life," detailing various forms of "moral insanity," including "affective insanity," applying to cases of disturbances in the "mode of feeling generally and not of moral feelings only" (cited in Walk 761). Subsequently, Karl Kahlbaum (1829–1899) described in 1887 what he believed to be three types of "partial pathology" that could affect children: paranoia, dysthymia, and diastrephia, which affected judgment, mood, and the will, respectively (Stone 83). These formulations of mild or partial mental illness in children countered old assumptions that children were largely immune from mental disorders, excepting idiocy and fever-induced hallucinations and delusions. Moreover, these formulations helped alter prevailing understandings of mental illness as purely inborn by suggesting the mediating role of social environments, first formulated in relation to the contagion of adverse circumstances.

Drawing upon his work at one of America's first research universities, Johns Hopkins University, the psychologist Stanley Hall argued that children who lacked proper parental guidance and education risked mental illness in adulthood. Adult pathology and/or deviance could be explained by developmental failures. Hall was particularly concerned by failures of development during "adolescence," a new space of visibility carved out by his developmental ontology. Hall emphasized that adolescents are particularly vulnerable to corruption (Finn 93). Synthesizing contagion with heredity, Hall viewed lower-class youths as particularly susceptible to corruption because of their inherent, class-based dangerousness. He also viewed lower-class youths as a source of contagion for corrupting higher-class peers (Finn 93). Surveillance and control were required strategies for containing working-class adolescents' potential dangerousness. In contrast, Hall viewed middle-class adolescents as innocent; their successful transition to adulthood required protection from contamination and cultivation of self-discipline (Finn 98).

Childhood was being constituted by educational, hygienic, and medical professional discourses as fraught with increasing peril. Yet the emerging medical, psychological, and educational professionals who described these perils offered practical insights and disciplines that would aid parents, teachers, and helping authorities/philanthropists. For instance, Hermann Emminghaus (1845–1904) encouraged collaboration between pediatric medicine and psychiatry to address particular forms of childhood psychoses that he believed stemmed from adverse circumstances (Walk 754–67). Emminghaus helped legitimize the idea that future social pathology might be prevented by ameliorating adverse social circumstances.

The idea that adverse social conditions might contribute to social pathology and future delinquency by shaping children's character drew upon and transmuted nineteenth-century discourses of motherhood, further contributing to the insertion of mothers within medical practice. The mother, previously implicated in producing her children's morality, was by the beginning of the twentieth century directly implicated in childhood mortality, delinquency, degeneracy, and madness. Consequently, the private space of the family required more expert surveillance and intervention in order to educate mothers, particularly immigrant mothers, and to target children at risk for various forms of dangerousness. Settlement house workers, day nurseries, visiting nurses, and child-rearing workers drawn first from the ranks of private philanthropy (but eventually professionalized and supported by the state) sought to educate working-class, immigrant mothers on proper feeding, hygiene, and child-rearing standards. In the case of juvenile delinquency, this complex sought to "forestall the drama of police action by replacing the secular arm of the law with the extended hand of the educator" (Donzelot 97).

At the turn of the twentieth century, child advocates promoted expansion of public education to prevent future delinquency and mental degradation. An 1893 text authored by Sarah Cooper, who helped found California's kindergarten system, asserts:

> To start from the very foundation of things, we are compelled to admit that a large proportion of the unfortunate children that go to make up the great army of criminals, paupers, and lunatics are not born right. They come into the world freighted down with evil propensities and vicious tendencies. They start out handicapped in the race of life" due to the "great, divine, inexorable law of heredity." (89)

Despite this great pessimism, the text expresses optimism in the role of "infant" education, particularly "character" education, in combating degraded parental and hereditary influences. Similarly, Lyman P. Alden writes in 1885 that the majority of children receiving public assistance have "inherited tendencies to wrongdoing more or less marked" or have "acquired habits" making them unfit. Yet Alden expresses optimism in the capacity for "skillful training and considerable time" to "eradicate" bad habits and "build up new character" (cited in Reef *Childhood* 159–60). Education thus assumed vital importance as a security technology within the child-saving discourse.

By 1900, thirty-two states in America had compulsory education, yet census data from 1900 found that nearly 20 percent of children between the ages of ten and fifteen were part of the paid labor force (Reef *Working* 182). Child labor, while still legal, was becoming subject to public critique and was no longer uniformly regarded as an end unto itself. Photographic images of child labor produced by Lewis Hine between 1907 and 1918 bolstered popular support for child labor restrictions (Reef *Childhood* 155). Child-saving activists claimed child labor kept wages low and the poor downtrodden. Perhaps more importantly, child laborers were denied the remedial childhood socialization offered through public education. This remedial education for the "dangerous classes" emphasized inculcation of normative values, including those reinforcing gendered and class distinctions (Pollock xviii). Child savers and policymakers believed that the "unfortunate classes'" lack of the "foundations for self-government" could be combated by incorporating values into the educational curriculum, particularly the "principles of morality, thrift, industry, and self-reliance" (cited in Preston and Haines 32). Education therefore emerged as a technology for transforming poor children into respectable citizens. Thus, the productive value of children's labor was becoming overshadowed by the dangers child laborers posed to national security. However, these concerns about child labor did not gain wide enough currency to overcome laissez-faire regulatory rulings until the depths of the Great Depression of the 1930s.

THE TWENTIETH-CENTURY "TUTELARY COMPLEX"

Many fathers and mothers in the United States were unable to provide for their children during the early decades of the twentieth century despite punitive tramp laws and philanthropic or public assistance. In urban areas, families lived in crowded tenements, even after the passage of housing laws designed to prevent crowding (Reef *Poverty* 109). Many of these families were recent immigrants: 9.4 million people entered the United States between 1906 and 1915, despite efforts by the United States to restrict immigration, particularly by "undesirables" such as those with epilepsy, the insane, prostitutes, and anarchists (110). Later immigration restrictions passed in 1917, 1921, and 1924 tried to limit entry by country of origin, literacy, health, morals, political beliefs, and so on (110). Despite these increasingly restrictive immigration policies, the urban areas grew in population density, and the factories, many of them sweatshops, offered workers poor wages and working conditions. When the U.S. government began collecting data on income levels and poverty in 1915, analysts concluded that 78 percent of U.S. fathers earned less than the $700 a year estimated as necessary for sustaining an average family (calculated as consisting of 5.6 people) (112). Progressive reforms aimed at saving children were inadequate for redressing the extent of hunger and deprivation characterizing the lives of many poor households. Persistent popular sentiment that the poor remained poor because of their own indolence prevented changes in work and remuneration practices that blighted children's lives.

Although Western children's lives did not change significantly with the turn of the century, academic and social interest in the characteristics of these lives grew, eventually encouraging the institutionalization of state apparatuses aimed at improving childhood education and health. Early twentieth-century medical, philanthropic, and public interest in children's welfare led to the creation of mothers' pensions and the institutionalization of specialized public and private institutions dedicated to child welfare and guidance. World events, including World War I and the Great Depression of the 1930s, encouraged greater interest in children's future capacities to contribute to the security and economic productivity of nations. Economic and political changes in the 1930s and 1940s would produce social welfare apparatuses and greater legislative protections in the United States aimed at protecting children as vulnerable citizens in need of careful surveillance and guidance.

At first, the concern about children as a social policy issue was fueled by eugenic concerns about the health of the population and threats to its moral vitality, including delinquency and heritable degeneracy. As Nikolas Rose explains in *The Politics of Life Itself*, in the late nineteenth and early twentieth centuries, "the nation was not only a political entity, it was a biological one. It could be strengthened only by attention to the individual and collective biological bodies of those who constituted it" (138). The nineteenth-century medical-hygienist experts and philanthropic authorities sought to contain threats through enclosure in disciplinary spaces such as orphanages. Enclosure created opportunities for controlling behavior, thereby producing the socializing practices of the tutelary complex, which aimed at the moral education and reformation of potentially dangerous children. These practices also circulated throughout society as mothers were encouraged to adopt prescribed sanitary and child-rearing practices. As the philanthropic complex underwent professionalization and as the social sciences proliferated, expert authorities expanded their attention to include understanding and profiling the "natural development" of all children (Singer 68). Child studies subjected children to the measurement devices of scientific inquiry: children thus became the object for the collection of "empirical" data about "natural" developmental processes (see Rose *Governing* 123–216; Morss 11–48). Newly created knowledge about the nature of childhood was directed beyond diagnosis and therapy as it was instrumentalized in the *prevention* of mental disorders and criminal behavior through ever more refined instruments designed to detect pre-madness and pre-delinquency (e.g., see Bridges 531–80). These preventive knowledges and technologies operated as security mechanisms whose circulation slowly supplemented, and at times replaced, nineteenth-century disciplinary regimes. By the 1920s, child saving had expanded its focus beyond subnormal, dependent, and delinquent children to include "normal" children, as child savers emphasized strategies for adapting children to societal norms while simultaneously attempting to understand the stages of a "normal" child's development. The goals of socialization thus extended beyond segregation, rehabilitation, and prevention of deviance in poor children to incorporate normalization of *all* children.

Early Twentieth-Century Biopolitics of Childhood

Roger A. Deacon describes how the rise of the academic and professional helping disciplines drew upon and transformed Christian confessional practices in the course of rendering family life visible and subject to intervention. Although disciplinary methods that divided space and regulated behavior and thought had been in place for centuries in monasteries, armies, and workshops, eighteenth- and nineteenth-century fears of social and medical contagion opened up the family in new ways to expert and popular surveillance and normalizing disciplines designed to reduce these threats (see Foucault *History* 108–112). By the early twentieth century, the family home and school were subjected to what Foucault describes as a kind of "cellularization" of space and time that normalized comportment and facilitated surveillance by parents and by expert authorities concerned with family welfare (Deacon "Truth, Power" 448). Hierarchical observations and normalizing judgments were derived from observations of cellularized spaces. Parents were encouraged to monitor their children's sleep and daily routines to study behavior for secret vices or suspect tendencies. Suspicious behaviors could be reported to child experts who would compare a particular child's comportment with newly created norms of behavior. Expert authorities also infiltrated schools with the objective of identifying future deviants through the application of newly developed childhood screening tools.

Treatment and prevention motivated enhanced surveillance and detection by both parents and expert authorities in the early twentieth century. The idea that children were malleable appealed to public sentiment because it suggested that social stability and democracy could be engineered. The authorities writing in the early 1900s on childhood were often quite explicit about the rationale for child guidance. For instance, in 1900 Charlotte Perkins Gilman's *Concerning Children* links effective, reasonable, and nonpunitive discipline to "wise, strong, self-governing citizens" (95). Ellen Key's 1909 *Century of the Child* delineates children's rights in the context of human progress, realized through the careful government of the emancipated child. Gilman and Key were concerned about children because they saw childhood socialization as critical to democracy and national well-being. What follows briefly explores the new institutions, knowledges, and measures used to represent, delineate, and transform childhood in the early twentieth century.

As explained in the last section, the process of rendering the family space visible in the late nineteenth century began with the referral of delinquent or dependent children to the juvenile courts. Establishment of the legal doctrine of parens patriae, or "the State as parent," afforded the state the right to force families to accept visitation by social workers and enabled judges to refer children to special clinics established early in the twentieth century. Among the first of these clinics was the Lightner Witmer's Psychological Clinic established in 1896 at the University of Pennsylvania and the Chicago Juvenile Psychopathic Institute founded in 1909 (Stevenson 17). In his review of these clinics' formation, Stevenson explains that the Chicago clinic was the first children's clinic to combine "psychiatric, psychological, and social approaches" to explaining and rehabilitating childhood deviance (17). This clinic was headed by Dr. William Healy (1881–1962). Healy's 1915 text, *The Individual Delinquent*, was regarded as the authoritative text on childhood deviance.

Although many of the early childhood guidance clinics focused on delinquency, still other clinics focused directly on children's mental health. The Henry Phipps Psychiatric Clinic of Johns Hopkins Hospital addressed mental illness in children as part of their more general mental hygiene practice. The mental hygiene movement sought to treat and prevent mental illness through community clinics. Leo Kanner's (1894–1981) work with children at Johns Hopkins led to the creation of a new diagnostic category, autism, that would capture the attention of child psychiatry and psychology in the late twentieth century (see Nadesan *Constructing*). By the 1920s, child-oriented mental hygiene clinics often relied on state-sponsored social workers to implement or supervise treatment regimes for children suffering from their newly discovered susceptibility to mental illness in addition to their long-established susceptibility to delinquency.

As explained by Kathleen Jones in *Taming the Troublesome Child*, public interest in children's proclivity for delinquency or mental illness prompted careful professional delineation of normal behaviors, affective states, and cognitive capacities. Experts strove to establish the characteristics of "normal" stages of children's intellectual and social development. Arnold Gesell's (1880–1961) work in 1911 at the Yale Psycho-Clinic pursued an empirical approach to studying the "stages" of child development. At the Yale clinic, Gesell observed and treated children who were having problems at school and also studied the "normal" developmental patterns of infants and toddlers. Based on his research with these populations, Gesell created standardized developmental scales that, as summarized by Nikolas Rose in *Governing the Soul*,

> introduced a new division into the lives of small children, a division between normal and abnormal in the form of the differentiation of advanced and retarded.... Norms of posture and locomotion; of vocabulary, comprehension, and conversation; of personal habits, initiative, independence, and play could now be deployed in evaluation and diagnosis. (147)

These scales were widely disseminated within the medical community and became institutionalized as "objective" measures of "normal" development.

Creation of the standardized intelligence test in 1905 by Alfred Binet, Victor Henri, and Theodore Simon contributed to efforts to classify children according to "scientifically" derived measurements of normal and abnormal "intelligence." Intelligence emerged in the early twentieth century as perhaps the most important factor seen as impacting susceptibility to crime and other forms of degeneracy. Hence, expert authorities struggled to create ever more refined measures for identifying children with "low" intelligence. Accordingly, in 1910 Henry Herbert Goddard (1844–1957) refined Binet's and Simon's measures by introducing the category of the "moron" as a form of mental subnormality above that of idiocy (Trent 161). Goddard drew upon Mendelian models of heredity that had recently been popularized to argue for the heritability of mental deficiency and suggested a link between mental deficiency and criminality.

Goddard's formulation of the moron occurred immediately after publication of D. Collin Wells's essay, "Social Darwinism" (1907), which elaborated in great detail upon the dangers criminals, "defectives and dependents" posed to society (701). Wells and other eugenicists argued for sterilization and enforced institutionalization of those regarded as mentally defective (feebleminded, insane, and epileptic) to prevent their reproduction and purported contaminating influences. Goddard's new formulation expanded the class of potentially dangerous people by developing a more nuanced instrument for detecting mental variations. Goddard's pessimism regarding this class of individuals worked against child savers' logic of prevention. It was not until the late 1920s and early 1930s that education would triumph in the public imagination as an instrument of social progress capable of combating heritable inferiority

William Healy and Edith Spaulding helped undermine popular suspicion that nearly all immigrant children belonged to the moronic class. In 1914 Healy and Spaulding published an essay using a thousand cases of "young, repeated offenders" to debunk the idea that criminality was inborn

(837). Using the Binet-Simon intelligence tests, they also concluded that most immigrant children had "normal" intelligence. They suggested that environmental factors had to play an important role in explaining the delinquency of the children referred to Healy's Juvenile Psychopathic Institute (Richardson 79–80).

By the 1920s many new classes of child experts—pediatricians, child psychologists, and child psychiatrists—were creating and popularizing knowledge about normal and abnormal stages and processes of child development. Although these descriptions first emphasized cognition, they slowly embraced personality as well. In 1924 Jean Piaget's (1896–1980) *Judgment and Reasoning in the Child* articulated "normal" stages of childhood cognition, thereby enabling the identification of "abnormal" or delayed cognitive development. While working on the problem of mental hygiene at Johns Hopkins, in 1939 Leo Kanner published the first child psychiatry textbook in English, *Child Psychiatry*, which acquaints pediatricians with the personality problems of children. Appel and Strecker's *Practical Examination of Personality and Behavioral Disorders: Adults and Children* illustrates the systematic and "scientific" approach to studying and diagnosing multiplying forms of social pathology. Much of early twentieth-century research and texts focused not only on cognitive norms of development but also on the *development of normal affect and interpersonal relations in the context of family dynamics* and school relationships (see Jones 120–48).

Accordingly, David Armstrong concludes in *A New History of Identity* that the early twentieth century engendered a new approach to medical subjectivity centering on surveillance, mental hygiene, the child, and interpersonal dynamics (100–103). This approach would evolve over the course of the twentieth century as new dangers to the "psycho-social space of interpersonal hygiene" were discovered everywhere (154–55). As the definition and scope of what constituted a "health problem" expanded to include interpersonal hygiene and, eventually, mental hygiene, expert authorities turned to explore and manage the social-psychic environment of America's children. Stevenson reports that children brought to the attention of the clinics in the 1920s and 1930s were reported most frequently for the following conditions:

> disobedience, negativism, stubbornness, rebelliousness; nervousness; temper; stealing; truancy; lying; feeding difficulties; "Does not get along with other children"; retardation in school; Enuresis; school failure; speech difficulties; disturbing behavior in school; fingersucking and nailbiting; placement, adoption; overactivity; shyness, withdrawal; sleep disturbance; fears; excessive phantasy. (56)

Parents, relatives, courts, and social agencies all referred children to the clinics, but it was the schools that served as the primary sites for observing the psychic and interpersonal hygiene of America's children.

Previous institutionalized health screenings in public schools provided the opportunity for expert surveillance of children's mental, as well as physical,

health. Just as the schools had served as laboratories for observing and intervening in the relationships between hygiene and germ-based disease, now
also the schools became laboratories for the application of the emerging
principles of mental hygiene. Heightened surveillance of the schoolchildren's
psyches engendered ever more psychiatric inquiry into childhood-based
forms of "deviance." By the mid-1920s, the child guidance clinics and
school officials shifted their focus from lower-class delinquents to the so-
called problem children of the middle class (Jones 120–48). Increasingly,
interpersonal skills and relationships in the home and at school were subject
to surveillance so that the factors "motivating" the child's delinquent behaviors could be detected.

The effect of this focus on developmental perils was that the "development"
of "normal" middle-class children gradually came to be regarded as fraught
with peril and as in need of careful study and intervention. Evaluations of
children's social, psychic, behavioral, and cognitive performance were made
in relation to the characteristics of the "normal population" rather than
in relation to the stigmata of degeneracy or the idealizations of perfection
(Armstrong "The Rise" 397). As a widening range of psychological neuroses and personality disorders were articulated in psychiatric vocabularies,
educational psychologists, school health officials and parents were instructed
how to identify incipient disorders and dangerousness in their otherwise
"normal" children.

Responsible Motherhood and Early Twentieth-Century Childhood Normalization

Popular diffusion of expert knowledge about newly established "norms"
of children's affective and cognitive development slowly shaped the duties
and knowledge base of "responsible motherhood" (see Apple 1–154; Jones
174–204). Women freed from domestic production by the rise of industrialization and consumer culture took a new interest in motherhood and actively
sought out expert child rearing advice that advised them to assume personal
responsibility for development of their children's moral and psychic normality.
For example, *Parents Magazine*, established in 1926, was a well-circulated
outlet for child-guidance precepts. Within such popular media, middle-class
readers found that "the troublesome child acquired an 'everyday' face, one
that looked remarkably like the reader's daughter or son" (Jones 97). Mothers
were advised to monitor their children for symptoms of deviance and pre-
delinquency. In effect, mothers were responsibilized for their children's nor-
malcy. Abnormalities in child behavior or mental development risked not only
the health of the child but also the very strength of the nation.

Annette Dorey's work on the "better baby" of the 1920s and 1930s
illustrates how public health officials disseminated to mothers the emerging
expert knowledge about child development. Promoted by the American
Baby Health Contest Association, the organizers and promoters hoped that
the contests would minimize infant deaths, foster maternal responsibility,

and help "secure and compile accurate information which will assist parents, educators and scientists to better understand the laws of Child Development" (cited in Dorey 69). Unlike the "pretty baby" contests of former days, these contests attempted to gather and standardize information about "perfect babies." Perfection was defined by the babies' "health," which was determined in relation to norms of physical growth as well as norms of mental development, both of which were assessed with scorecards. Mental development was primarily operationalized in terms of simple behavioral standards (e.g., "sits alone" and "plays with simple objects") and disposition factors (e.g., "irritability," "highly nervous," and "mentally balanced"), as little standardized criteria existed (cited in Dorey 46, 249). In effect, Dorey contends, the prevailing medical science viewed the child" as a collection of distinct parts that could be measured" (70) and evaluated in relation to emerging standards of perfection. Hereditary (linked to the eugenics movement) and domestic practices, which were understood in relation to sanitation, sleeping conditions, systematic feeding, and so on, were held to account for babies' standards of health. In the quest for better babies, that is for "a 'superior crop' of babies" (Dorey 5), mothers were provided with pamphlets and other literature on child hygiene and care including practices such as "Daily bathing," "Sufficient sleep and rest," and "Cleanliness and fresh air" (cited on 134). Thus, the normalization of American babies around "optimal" physical and disposition standards was accomplished in large part by standardizing and formalizing "expert" knowledge about what is now regarded as basic infant care, which was disseminated through popular periodicals, pamphlets, and other mediums (see also Apple 1–154).

In addition, Sarah Nettleton's work on dentistry illustrates how emerging expert knowledge of dentistry inscribed and prescribed standards of good mothering. Dentistry emerged in the nineteenth century as part of the medico-administrative apparatuses aimed at securing public health. "Dental diligence," or the knowledge required for the dental health care system, entailed a wide range of surveillance and therapeutic practices, including regular "dental examinations, epidemiological surveys, dental health education programmes and dental social research" (102). The school, the dentist's office, and the home served as the primary sites for dental surveillance and discipline. The mother emerged as the " 'natural' dental agent," responsible for the health and welfare of her child (103). Although mothers were regarded in the early twentieth century as intuitively responsive to their children's needs, they were also seen as requiring expert guidance on the proper care of their children, particularly where dentistry was concerned. Accordingly, dental health officials prepared guidelines on infant feeding and also successfully implemented school screening programs. These programs articulated the mouth as "the medium which permitted the dental gaze to focus upon the child's home" (103). Nettleton concludes that the practices and discourses of good dental care thus operated to produce mothers "as agents of dental government" (109). At stake was the health and vitality of the nation.

Behaviorist psychology of the 1920s offered parents another framework for governing children through regimented practices. As Peter Stearns explains in *Anxious Parents*, behaviorism focused on environmental manipulation as the key to parenting and emphasized rewards for good behaviors over discipline and punishment (66–67). Although behaviorism encouraged mothers to adopt highly disciplined regimes of child rearing, these regimes were productive in orientation rather than punitive and therefore operated preventively. Indeed, as Stearns recounts, the famous behaviorist John Watson was strongly against traditional punishments such as spankings because he felt that they were irrational and punitive devices. Although many of the behaviorist principles were never adopted widely, Stearns points out that the general goal to govern the children by manipulating their environment so as to render discipline unnecessary became an essential dimension of twentieth-century child rearing (71). This goal was consistent with new beliefs about children's vulnerability and required expert guidance.

Intensive mothering was not the only new role assumed by women in the early twentieth century. More affluent women were also invited by advertising to see themselves as consumers. Newly created products and services promised women enhancement of their social mobility, interpersonal relationships, and domestic achievements. As Stuart Ewen's *All Consuming Images: The Politics of Style in Contemporary Culture* demonstrates, advertising was a pervasive form of communication that circulated throughout the social body, cultivating new technologies of the self that were viewed as commensurate with the values of democratic citizenship and the pursuit of health, happiness, and economic success.

Yet advertising messages were not without contradictions. Much of the early advertising and consumer research produced the consumer psyche as a feminine space ruled largely by instinctual and irrational impulses (Graham 539–65). Still other advertising discourses articulated the female consumer as a rational, home engineer. Laurel Graham describes how civic organizations, women's clubs, and reformers combated the advertiser's model of irrational personhood through their efforts to educate women in rational consumption. Accordingly, Graham chronicles the emergence of a new kind of expert, the "household engineer," as illustrated by Lillian Gilbreth (1878–1972), who published *The Home-Maker and Her Job* in 1927. In addition to educating women about rational consumption practices, Gilbreth offered strategies for household management aimed at enhancing domestic cooperation and efficiency. Gilbreth argued that effective household management required women to assume the roles of amateur psychologist and engineer. The "responsible mother" used psychology to study each family member's needs in order to best allocate household duties in a manner ensuring cooperation. Through her analysis, Graham concludes that Lillian Gilbreth "imagined a society in which women were managed through rather than against their own autonomy" (554). Management through autonomy entailed technologies of the self, or practices of ethical self-formation, whereupon women produced themselves as rational, self-governing household agents. These early practices of interpellating women—as rational, self-governing household agents and responsible mothers, and also as irrational, impulse-driven consumers— would become more important in the post–World War II era, as will be discussed later in this chapter.

CHILDREN AND TWENTIETH-CENTURY SOCIAL-WELFARE

In general, however, the growth of advertising and consumer culture did not profoundly alter the lives of many Americans in the early decades of the twentieth century. Many people remained mired in absolute poverty and lacked the resources to purchase adequate food let alone the more frivolous or discretionary consumer commodities. As explained in the context of Grace Abbott's 1932 essay on child social welfare reforms, child poverty and child labor remained stubbornly high throughout the early decades of the century. The next section explains how the persistence and exacerbation of high rates of poverty, particularly child poverty, ushered in social welfare reforms during the 1930s that shifted risk away from individuals toward collectivized government apparatuses. The "helping hand" ethos of child government that would emerge out of this social welfare logic would eventually produce a wide range of child-welfare programs in the 1960s and 1970s (DeMause 52). However, as will be explained, these programs often assumed that child deprivation was the result of a culture of poverty.

Instituting Social Welfare Logics

Early in the twentieth century, some Progressive reformers in the United States advocated for the expansion of public support for poor women and children because of the inadequacy of private philanthropy. At the same time, industrial workers agitated for better wages and working conditions. Neither reformers nor laborers were particularly successful in the first couple of decades of the twentieth century in legislating effective reforms for workers. Child poverty therefore remained a pressing problem in need of solution given its linkages to state security.

One early effort to reduce child poverty was the public institutionalization of outdoor relief for "deserving" poor women. Between 1911 and 1921, forty of the country's forty-eight states passed mothers' pension laws that provided public cash payments to allow poor mothers to raise their young children at home. This system coexisted with the orphanage system. However, these pensions were inadequate to basic survival and were provided only to a very small portion of the population (Sealander 101–102). At the same time, "tramp laws" (104) were passed to make it a crime to be a man without visible means of support. As Sealander summarizes: "Mothers' pensions—linked to criminalization of a husband's nonsupport of dependents—reflected outdoor relief's guiding principle: alms should never be sufficient for comfort, lest they promote sloth" (Sealander 105).

The need for public assistance to combat dire poverty stemmed in large part from the poor work conditions that prevailed throughout the first three decades of the twentieth century. In the United States, a Supreme Court hostile to labor continued to hand down decisions reinforcing market autonomy (over worker rights) and limiting the capacity of the state to regulate market transactions, particularly in the context of the relationship between employer and employee. The industrial and agricultural surplus labor supply produced by immigration guaranteed that wages were kept low and work conditions poor. For instance, Henry Ford's five-dollar-a-day wage for workers in 1914 was exceptional for the time. Early twentieth-century trade unions, such as the International Ladies Garment Workers in 1900, were crippled by the antitrust provisions of the 1890 Sherman Anti-Trust Act, until passage in 1914 of the Clayton Anti-Trust Act, which allowed some strikes by exempting them from antitrust regulations. Still, industrial workers were not guaranteed the right to collective bargaining until passage of the Wagner Act in 1935 under Franklin D. Roosevelt.

The inhospitable environment to labor prevailing in the early part of the twentieth century adversely impacted family welfare, according to social welfare advocates. Accordingly, in 1932 Grace Abbott lists parental unemployment and "low wages" as universal "enemies of childhood" and cites with approval a quote by Julia Lathrop, founder of the Child Bureau: "The power to maintain a decent family living standard is the primary essential of child welfare" (cited in Abbott XX4). Franklin D. Roosevelt's New Deal reforms beginning in 1933 and instituted throughout the 1930s introduced labor protections and expanded social welfare apparatuses significantly. For instance, the *National Industrial Recovery Act* of 1933 aimed to increase employment, improve housing for the poor, demonstrate to private industry the viability of public works projects, and rehabilitate inner-city slum areas (Jackson 220). The *Wagner Act* of 1935 gave industrial workers the right to collective bargaining, while the 1938 the *Fair Labor Standards Act* set a minimum wage for workers involved in interstate commerce and outlawed most forms of child labor (Reef *Poverty* 152).

The 1930s and 1940s were a unique period in U.S. history because economic circumstances and political events converged in such a way that policy makers publicly acknowledged that U.S. economic markets failed to produce enough jobs to employ Americans in jobs that paid above poverty wages. This public acknowledgment constituted a repudiation of laissez-faire logics that regarded the unregulated market as necessarily rewarding individual industriousness. As Foucault explains, the (social welfare) liberal logics that emerged at this unique historical juncture held that preservation of democratic freedoms required economic interventionism (*BoB* 68). Public policy responses therefore expanded state apparatuses to compensate for market failures to meet citizen needs and enabled workers to voice demands. Policy also aimed to protect financial capital from excessive risks by regulating financial institutions (e.g., *Glass-Steagall Act* of 1933 and the *Securities Act* of 1933). The scale and impact of New Deal reforms were denounced by many critics as threatening freedom by constraining its most vital operations, market transactions.

The expansion of government apparatuses to be discussed presently introduced collective forms of risk management for the nation's poorest populations, for the elderly, for poor children, and for some disabled populations. In 1935, in the depths of economic depression, the United States passed the *Social Security Act* (Act of August 14, 1935) [H. R. 7260],

> An act to provide for the general welfare by establishing a system of Federal old-age benefits, and by enabling the several States to make more adequate provision for aged persons, blind persons, dependent and crippled children, maternal and child welfare, public health, and the administration of their unemployment compensation laws; to establish a Social Security Board; to raise revenue; and for other purposes. ("History of Social Security")

Title IV of the act provided grants to states for Aid to Dependent Children (ADC) (later renamed Aid to Families with Dependent Children [AFDC] in 1960):

> SECTION 401. For the purpose of enabling each State to furnish financial assistance, as far as practicable under the conditions in such State, to needy dependent children, there is hereby authorized to be appropriated for the fiscal year ending June 30, 1936, the sum of $24,750,000, and there is hereby authorized to be appropriated for each fiscal year thereafter a sum sufficient to carry out the purposes of this title. The sums made available under this section shall be used for making payments to States which have submitted, and had approved by the Board, State plans for aid to dependent children.

The act stipulated state administration of the grants. Most aid to children in the 1930s and 1940s went to white, widowed mothers (Reef *Poverty* 177). Additional titles to the act included aid to the blind, increased appropriations for the Public Health Services, vocational rehabilitation, and infant and maternal health (Katz 245). Still, many Americans remained very poor during the 1940s and were ineligible for, or simply unable to get, provisions under the Social Security Act (e.g., ethnic minorities).

Growing public concern for the welfare of children in the mid-twentieth century encouraged development of more programs targeted specifically at their security. Children's welfare mattered because it was becoming ever more closely linked to state security. For example, the National School Lunch Act of 1946 subsidized states for food and equipment costs associated with providing a nutritious lunch for low-income children. Section 2 of the act defines its purpose in relation to national security:

> It is hereby declared to be the policy of Congress, as a measure of national security, to safeguard the health and well-being of the Nation's children and to encourage the domestic consumption of nutritious agricultural commodities and other food, by assisting the States, through grants-in aid and other means,

in providing an adequate supply of food and other facilities for the establishment, maintenance, operation and expansion of nonprofit school lunch programs. (U.S. Department of Agriculture, "National School Lunch Program)

In 1953 the U.S. Department of Health, Education, and Welfare was established, and in 1954 the School Milk Program subsidized milk for eligible school children (Reef *Poverty* 175).

In the post–World War II era, social welfare advocates argued for extension of government social security programs while promoting Keynesian, demand-focused economic policies. The "Fordist" economic regime that thrived in the 1950s and 1960s was made possible in the postwar landscape by lack of global competition and cheap oil. Keynesian-inspired government spending encouraged a military industrial complex on the one hand and social programs aimed at securing old age, bolstering education, and providing health services for the young on the other. Minimal welfare provisions were made available for the nation's poorest populations. In 1965 Medicare and Medicaid programs were formalized. Head Start was also created in 1965. In 1979 the Department of Education was split off from the newly titled U.S. Department of Health and Human Services (see http://www.hhs.gov/about/hhshist.html). Having now examined the shift in risk enacted by these social welfare programs, discussion turns to examine the evolving logic of childhood government emerging in concert with these governmental policies.

The Helping Hand and the Therapeutic Ethos

Strategies for governing U.S. children in the second half of the twentieth century were inflected by region, socioeconomics, and ethnicity, among a variety of other factors. Still, despite considerable heterogeneity, many cultural observers point to the growth of a therapeutic ethos that increasingly shaped the adult-child relationship. Although this therapeutic ethos was applied to the children of the lower socioeconomic classes (e.g., as with Head Start), it held greatest relevance for the child-rearing attitudes and practices of the solidly middle and upper classes. Remarkably, however, the helping hand was extended to childhood populations previously marginalized from public concern, particularly disabled children.

DeMause's "helping hand" mode of childhood socialization focuses primarily on the view of child rearing that emerged in the mid-twentieth century out of psychoanalytic and humanistic psychology:

> The helping mode involves the proposition that the child knows better than the parents what it needs at each stage of its life, and fully involves both parents in the child's needs as they work to empathize with and fulfill its expanding and particular needs. There is no attempt at all to discipline or form "habits." (52)

Roger Cox explains that newborns were understood as having inherent souls requiring freedom to "develop" on their "own terms" (190). Echoing earlier romanticist formulations of childhood education, the helping-hand model encourages parents and professionals to steer each child in the form of an "invisible pedagogy" (Cox 191).

The idea of an "invisible pedagogy" may seem to suggest less invasive forms of child rearing, yet what is masked is how the child's interiority is revealed as a space requiring surveillance and control (see Rose *Governing* 123–205). Thus, although the helping-hand model lacks the forceful authoritarianism of times past, it substitutes direct control with a panopticon of invisible pedagogies aimed at producing the subjects of liberal democracy. Guiding these invisible pedagogies of domestic life are the expert knowledges and therapeutic practices engineered by privately and publicly funded "experts" of the helping professions, including child psychiatry, child psychology, and educational psychology. The experts populating these fields helped expand the apparatuses of the liberal welfare state. They also instructed individuals in the technologies of the self that promised to enable self-government (see Rose *Governing*).

Although the helping-hand mode of child rearing, complete with its invisible pedagogies, reflects assumptions and practices specific to the 1960s Human Potential Movement, its earlier roots can be traced to the rise of psychoanalytic theories and experts shaping early twentieth-century expert knowledge about family life. In the 1940s and 1950s, psychoanalytic principles were disseminated through popular culture in advice columns, baby books, and other expressions of child guidance. In 1952 the American Academy of Child Psychiatry was established (Miller and Leger 17). Popular circulation and public acceptance of psychoanalytic principles produced an imposing biopolitical matrix. New authorities claimed privileged knowledge of the formation of, and threats to, the young child's developing ego in the post–World War II context. Although divergences in approaches existed, psychoanalysis understood the infant as engaged, from the moment of its birth, in a fierce struggle for consciousness: for the development of its "ego." In the infant's phantasy and in "concrete reality," the comforting maternal presence anchored the infant as he or she struggled through the perilous seas of unconscious drives and impulses (Klein 130–143). Maternal absence or ambivalence (in phantasy or "reality") posed grave risks to the infant's future mental health, including the risks of psychoses, neuroses, and personality disorders. Psychoanalytical practitioners and advice columnists offered mothers advice on how to assist their child's struggle for integrating ego, superego, and id.

Mothers' willingness to look to expert advice had been cultivated by the proliferation of child guidance literature disseminated throughout popular culture in the early part of the twentieth century. Women pushed out of the workforce in the postwar environment found in child rearing a socially legitimate outlet for forging self and contributing to society (see Apple 83–134). Child rearing had become a critical part of the larger project of social engineering. Middle-class women's relatively isolated existence in their homes and suburbs may have made them particularly susceptible to "expert" knowledge developed by psychoanalytically informed experts emphasizing their importance in child development. Accordingly, as the work of psychodynamic theorists such as Anna Freud (1895–1982) and Melanie Klein (1882–1960) slowly infused child psychiatry, the role of parenting—particularly mothering—was seen as vital to the child's psychic development and personality, the latter of which was primarily defined in relation to an ability to initiate and maintain social relations. By the time humanistic psychology made inroads into popular culture, women were prepped to accept the personal and child rearing advice of psychological experts. Even apparently "normal" children benefited when their mothers employed expert advice in the course of their child rearing. Well-governed children contributed to the nation's well-being through their capacities for economic productivity and democratic citizenship.

In the 1960s and 1970s, humanistic psychology and psychoanalysis were fused in the "helping hand" mode of child rearing, which sought to create supportive interpersonal and group systems enabling and encouraging the developing child to learn self-regulation while achieving healthy personal development and moving toward self-actualization. Parents, teachers, and therapists were admonished to develop cooperative and nonauthoritarian relationships with their respective charges. Their primary purpose was to guide development and self-transformation rather than to prescribe it. As will be explained further in chapter 3, the "helping hand" mode of child rearing described by DeMause took on new inflections toward the end of the twentieth century with the rise of the cognitive paradigm and cybernetics.

More broadly, the helping-hand mode of child rearing encouraged passage of legal protections for children, particularly for children with disabilities and those who had suffered owing to poverty, racial, ethnic, and gender discrimination. The helping-hand ethos informed the social welfare reforms discussed earlier, including the lunch programs and the institutionalization of AFDC. These programs were also infused with a biopolitical sense of the need to enhance the vitality of the nation by strengthening children's nutrition and ensuring their basic needs. What was exceptional about the extension of the helping hand in the 1950s and 1960s was its application to child populations that had in previous years been subject to gross discrimination and institutionalized. Both African American children and disabled children were taken up as vulnerable subjects requiring affirmative action after decades—indeed, centuries—of marginalization and discrimination.

In the 1950s and 1960s, a number of legal acts were passed aimed at improving opportunities for disabled children. These early acts primarily focused on training educators to specially educate children with disabilities. They demonstrate the "helping hand's" adaptation to the specialized needs of children formerly shunned by society. For instance, U.S. PL 85–926 passed in 1958 encouraged training for teachers of students with mental retardation. In 1961 the Teachers of the Deaf Act of 1961 (PL 87–276) trained instructional personnel for children who were deaf or hard of hearing. PL 88–164 of 1963 "expanded previous specific training programs to include training across all disability areas" (http://www.ed.gov/policy/speced/leg/idea/history.pdf). In 1965 states were provided with direct grant assistance from the U.S. federal government to help educate children with disabilities through the Elementary and Secondary Education Act (PL 89–10) and also with the State Schools Act (89–313). The Handicapped Children's Early Education Assistance Act of 1968 (PL 90–538) authorized support for early childhood education programs, and the Economic Opportunities Amendments of 1972 (PL 92–424) increased Head Start enrollment for young children with disabilities. Still, as of 1970, only 20 percent of U.S. children with disabilities were educated in public schools, and many states continued to have exclusionary laws. Finally, in 1975, the U.S. passed PL 94–142, the Education of All Handicapped Children Act, now codified as IDEA (Individuals with Disabilities Education Act). The act guaranteed that children aged three to twenty-one with disabilities received free, appropriate education. In 1986 PL 99–457 mandated that states provide children with disabilities programs and services from birth (U.S. Office of Special Educations Programs). Section 504 of the 1990 Americans with Disabilities Act also broadly prevents discrimination against any qualified individual with a disability by any program or activity receiving federal aid.

Disability rights activism occurred in the context of a much broader civil rights movement. This civil rights activism has deep roots, but for the purpose of this chapter it will be discussed primarily in relation to changing attitudes about the value and government of African American children specifically and minority children more generally. The helping-hand model of child government contributed to, and was appropriated by, civil rights activists. Application of this model of childhood to African American children was historically novel.

It is important to keep in mind that the nineteenth-century child savers and the twentieth- century child guidance movement primarily, if not exclusively, focused on *white* children. Marlon Riggs's Emmy-winning documentary *Ethnic Notions* illustrates how African American children were routinely subject to dehumanizing and extraordinarily violent characterizations in popular culture throughout much of the early twentieth century. Popular circulation of these characterizations reinforced old sentiments that African Americans lacked the intelligence and virtues of white populations. Separate but equal educational doctrines were naturalized in many parts of the United States, and educated blacks were regarded as "uppity" and often ridiculed. Therefore, the 1954 Supreme Court decision in *Oliver L. Brown et al. v. The Board of Education of Topeka*, which found separate but equal unconstitutional, created the potential for revolutionizing educational opportunities for minority children. Unfortunately, the promise of this ruling remains unfulfilled at the turn of the twenty-first century.

Despite a real failure to desegregate many school districts in the United States, the civil rights movement did result in significant institutional reforms that broadened legal and economic opportunities for African Americans and other ethnic minority populations. The U.S. Civil Rights Act of 1957 established a Civil Rights Section of the Justice Department aimed at enforcement. Subsequently, the Civil Rights Act of 1964 prohibited racial segregation in schools, public places, and employment. The resulting efforts to enforce racial desegregation of schools in the southern states had to be supported by U.S. federal marshals, as dramatized by the poignant tale of Ruby Bridges. In 1960, six-year-old Ruby became the first black child to attend a segregated all-white elementary school. For months, Ruby walked to the school guarded by federal marshals who protected her from threatening white crowds (see Cole 1–24). Ruby's story illustrates both individual courage and the role of the helping hand in eroding nineteenth- and twentieth-century disciplinary spaces designed to segregate dangerous or "deficient" populations.

In the 1960s President Lyndon Johnson saw civil rights and poverty as intertwined, thereby legitimizing a new problem-solution frame linking social and economic justice. His "War on Poverty" involved the Economic Opportunity Act of 1964, which created an Office of Economic Opportunity (OEO) to administer a variety of community-based antipoverty programs. Yet the aims and strategies embedded in Johnson's programs inscribed elements of a culturalist account of poverty by focusing on job *training* for young, disadvantaged African Americans, rather than redressing the unavailability of well-paying jobs accessible to inner-city populations (Bremner 87). This reliance on culturalist explanations for poverty had the effect of reinforcing public perceptions that poverty was caused by a lack of personal initiative rather than by a lack of economic opportunity. This perception would ulti-mately encourage popularization of the "neo" logics of government, under-mining social welfare approaches and decoupling linkages between social and economic justice.

The Culture of Poverty: Reinventing the Dangerous Classes

By the end of the twentieth century, middle-class Americans had become increasingly distant geographically and culturally from working-class urban America. Poverty, particularly child poverty, was troubling to middle-class America, but understanding of the complex reasons why poverty occurred was vague. Public sentiment often converged around a vague sense that the poor were poor because they were improvident and lacked the skills or the motivation for economic self-sufficiency. This centuries-old attitude eroded public support for social welfare programs that offered minimal relief but failed to employ recipients. Middle- and upper-class attitudes about the causes of poverty, particularly urban poverty, were rarely tested by actual confrontations with *real* poor people because of the growth of suburbia in the post–World War II landscape.

The post–World War II transformation in urban landscape would have profound implications for the government of children, contributing to the eventual rebirth of the image of the dangerous, pre-delinquent. The spa-tial divisions that were produced by the post–World War II growth of the suburbs encouraged technologies and discourses that reinforced old binary distinctions between the deserving and the undeserving poor. Accordingly, even while social welfare logics and programs were promoted by politicians and social advocates, other discourses, technologies, and policies encouraged class and racial polarization. Neoliberal and conservative economic and cultural "reforms" would gain popular potency from this polarization the deserving and the undeserving poor.

As chronicled by Kenneth T. Jackson in *Crabgrass Frontier: The Suburbanization of the United States,* the Federal Highway Acts of 1916 and 1956, federal tax subsidies, and the Federal Housing Administration (FHA) (191, 203) instituted infrastructures and policies that enabled the "suburban dream" for an aspiring, white middle class who moved out of the cities into suburbia (190–218). FHA policies encouraged home builders to develop middle-class suburbs and ignore the minority and inner-city housing market (Jackson 213). In addition, by declaring whole urban areas ineligible for loan guarantees, the FHA caused the value of homes in many urban areas to decline, compounding the decline in value resulting from white suburban flight (213). This further discriminated against minority populations and institutionalized middle-class white privilege.

Inner-city slum areas were targeted for public housing projects, concentrating the "poor in the central cities" and reinforcing "the image of suburbia as a place of refuge from the social pathologies of the disadvantaged" (Jackson 227). Economic opportunities for inner-city residents declined as economic shifts and suburban flight limited job prospects for those without college degrees (W. Wilson 93, 97). As argued by William Julius Wilson, structural economic factors produced and exacerbated inner-city urban poverty, particularly for African American populations, but these complex economic predicates were rendered invisible in culturalist accounts of poverty that circulated in popular discourses.

In the 1960s Oscar Lewis wrote an essay titled "The Culture of Poverty" (19–25). His authority as an anthropologist helped legitimize his findings, but the popularity of the essay's thesis stemmed from its resonance with centuries of popular belief that the poor suffered from a degeneracy of lifestyle. Also in 1963, Gunnar Myrdal coined the phrase "under-class" for the "unemployed, and gradually unemployable persons and families at the bottom" of American society, cautioning that children of the underclass tend to remain entrenched in poverty (quoted in Bremner 86). Culturalist accounts of poverty stressing the degenerate living conditions of the poor and emphasizing the ways whereby the poor inhibit their own social advancement captivated the popular imagination. Children were targeted for reform just as they had been in previous eras. Yet, reform focused less on redressing economic conditions such as the unavailability of well-paying work than it did on saving deprived children from their families' culture of poverty.

Daniel Patrick Moynihan's 1965 report, *The Negro American Family: The Case for National Action,* denounced a crisis in African American families *attributable to their dysfunctional family culture:*

> The fundamental problem, in which this is most clearly the case, is that of family structure. The evidence—not final, but powerfully persuasive—is that the Negro family in the urban ghettos is crumbling. A middle class group has managed to save itself, but for vast numbers of the unskilled, poorly educated city working class the fabric of conventional social relationships has all but disintegrated. There are indications that the situation may have been arrested in the past few years, but the general post war trend is unmistakable. So long as this situation persists, the cycle of poverty and disadvantage will continue to repeat itself.

Moynihan concludes by quoting Gunnar Myrdal, who presented what he called America's greatest social problem: *"America is free to choose whether the Negro shall remain her liability or become her opportunity."*

In *Why America Lost the War on Poverty*, Frank Stricker argues that Moynihan fully acknowledged the role of economic factors in producing the "culture of poverty" he so vividly described (87). However, Moynihan's economism was deemphasized in popular accounts of his ideas that focused on his highly charged account of the Negro family. For middle-class Americans distanced from urban deindustrialization and blight, Moynihan's description of ghetto life resonated with televised and cinematic images negatively characterizing poor populations, particularly poor minority populations. While some sympathy and support existed for "disadvantaged" children, this support was limited in scope and application. Reports of rising individual and group violence among America's "under-class" (i.e., dangerous class) led some to consider breaking up the "ghetto family" in order to break the purported cycle of violence (Bremner 87).

Additionally, growing fears of juvenile delinquency may have tempered support for programs targeting poor children, as public perceptions of their nature oscillated between that of vulnerable victims needing a helping hand and that of risky subjects requiring discipline and incarceration. In the essay "Other People's Children," Robert Bremner describes how anthropological, sociological, and policy accounts of the 1960s and 1970s produced a spectacle of juvenile delinquency, promoting perceptions of the inherent dangerousness of the underclasses (85–87). This spectacle circulated in the popular media and vied with structural economic and prejudice-focused accounts of poverty and social marginalization for popular appeal. While the humanistic ethos of the helping hand may have helped soften public attitudes toward the children of these "dangerous" classes, the children's designation as inherently flawed would continue to inhere in the popular consciousness, awaiting restoration as *the* central representational paradigm in the late 1980s and 1990s. Chapters 3 and 4 explore these developments in further detail in the context of emerging economic and biopolitical formations. Before turning to these developments, this chapter concludes by sketching the new logics of government that arose in the United States in the late twentieth century, focusing in particular on their approaches toward the government of families.

NEOGOVERNMENTALITIES, CHRISTIAN PASTORALISM, AND THE DANGEROUS CLASSES AT THE TURN OF THE TWENTY-FIRST CENTURY

The tensions inherent in the social welfare policies of the 1960s and the 1970s contributed to the ease with which they were later dismantled. The idea of the culture of poverty eroded belief that poverty derived from lack of economic opportunity and promoted the idea that the poor were responsible for their own circumstances. In 1969 Richard Nixon concluded that "the present welfare system has to be judged a colossal failure" (cited in Wessel "Changing" A10). Nixon's proposal for a Family Assistance Program was rejected by a tax-averse U.S. Congress. Automation in industry and outsourcing trends began to erode union jobs in the 1970s, slowly shrinking the supply of well-paying jobs for nonprofessional, nonmanagerial workers. The expansion of a service-based economy in the closing decades of the twentieth century did create demand for workers, but most of these jobs paid poorly and lacked collective representation (i.e., unions). Poverty rates began to rise by the early 1980s at the same time that the incoming president Ronald Reagan decried "welfare queens" (cited in Wessel "Changing" A10). The convergence of negative attitudes about poverty and the economic demand for low-wage labor encouraged welfare reform. While advocating the ethos of personal responsibility, President Clinton formalized welfare reform with PL 104–193, the Personal Responsibility and Work Opportunity Reconciliation Act, in 1996. Clinton's reforms entailed creation of Temporary Assistance for Needy Families, or TANF, which replaced the previous program of federal assistance with a block-grant system for states. TANF required recipients to find work and established a five-year limit on benefits (Lueck A4). The government also raised the minimum wage from $3.15 an hour to $5.15 despite widespread resistance from business. President George W. Bush followed by offering an "Ownership Society" wherein citizens would assume more risk while reaping the promised benefits of personalized home ownership, control over retirement savings, tax credits or vouchers for education, job training, and health insurance (Calmes A1). The Ownership Society emphasizes "the individual, supplanting a 70-year-old approach in which citizens pool resources for the common good—and government doles out benefits" (A1). By 2002 welfare rolls had decreased by half, leading to celebratory claims of welfare reform's success (Lueck A4). Critics of welfare reform offer an alternative interpretation of the programs' effects. They claim that the reforms are punitive in effect and unjustly punish welfare recipients who are unable to secure employment. Moreover, critics claim that recipients who do find work are barely unable to sustain themselves on low hourly wages averaging $7.50 an hour (Schram 2).

Welfare reform and Bush's Ownership Society grew out of the three distinct, but interconnected, governmental logics that assumed social salience and political power beginning in the 1980s: neoliberalism, neoconservatism, and conservative Christian pastoralism. These movements diverge in many ways but share important common logics informing the government of children. They value market autonomy and encourage social welfare policy only insofar as that policy aims to encourage personal economic initiative. Employment is seen as the route to personal empowerment. Policy must reform the "culture" of poor parents to encourage personal industriousness and accountability (Ludwig and Mayer 176). Policy aiming at reducing intergenerational transfer of poverty tends to center on three broad areas: "schools, neighborhoods, and families" (177). What follows briefly summarizes contemporary neoliberal and conservative approaches to poverty alleviation focusing on these three broad areas.

Neoliberal, neoconservative, and conservative logics of government constitute families as central to the reproduction of dependency. Rather than addressing families' economic circumstances and opportunities, these logics of government emphasize family culture and parenting as integral to the reproduction of poverty. Single motherhood has emerged as a central problem-solution space for governing poverty in the United States (Kantrowitz and Wingert 46). Single motherhood emerges starkly in statistical representations created during the 1990s and early twenty-first century. During the 1990s almost 30 percent of all babies were born to unmarried women, leading to the concentration of poverty in mother-child families (Bianchi 307–333). In addition, owing to single motherhood and divorce, approximately half of the children born in the 1990s are expected to spend part of their childhood in single-parent homes (46). The U.S. Federal Interagency Forum on Child and Family Statistics reported in 2007 that according to 2005 data, 17 percent of children aged 0–17 lived in poverty, figured for a family of four at an annual income of $19,806; 21 percent of children lived in families with low income, defined as 100–199 percent of the poverty threshold; 32 percent of children lived in families of middle income, defined as 200–399 percent of the poverty threshold; 30 percent of children lived in the remaining high-income category (Federal 15). Many, but not all, of the 40 percent of children living in or "near poverty" live in single-parent households.

The problem space of single-motherhood has produced a vast array of preventive technologies and programs. However, while neoliberal strategies would simply adopt the most localized, market-friendly, and cost-efficient strategies for representing and redressing the "problem" of single motherhood, the rise of neoconservative and conservative governmental logics has significantly influenced government strategies. Accordingly, marriage promotion emerged as a central strategy for combating poverty despite critics' claims that marriage promotion fails to address the unavailability of well-paying jobs. President George W. Bush was particularly active in promoting this approach to poverty alleviation with his appointment of Dr. Wade Horn:

> As head of the Federal Administration for Children and Families, Dr. Horn
> has employed the zeal of an ideologue and the discipline of an academic to

inject marriage promotion into a host of government programs under his purview, even before Congress authorized an official marriage program. Today,
more than 200 programs are at work across the country, seeking to change
public attitudes surrounding marriage, persuade teenagers to aspire to matri-
mony and teach relationship skills to young couples. (Meckler A1)

Marriage-promotion programs are just one component of a culture-based
approach to combating poverty.

In order to prevent single parenthood, Christian charities, think tanks,
and policy groups persuaded federal and state officials across the United
States to promote abstinence through government-funded abstinence education programs and public service announcements. As government outsourcing and contracting grew during the 1990s and early twenty-first century,
Christian religious groups were able to secure public dollars for delivering
abstinence education. In 1991 Douglas Wilder, then governor of Virginia,
wrote in the *Wall Street Journal* that young black adolescents must abstain
from sex to save their families. His article criticizes a "total lack of discipline"
in black families headed by females and claims this lack as responsible for the
violence cited as the leading cause of death among blacks between ages 15 and
25. Hence, he argues that the inner-city "black family is teetering near the
abyss of self-destruction" while encouraging "our young—male and female
alike [to]—embrace the ultimate precaution—abstinence" (Wilder A14). By
1992 students everywhere began to be admonished in public schools to take
"chastity vows" (Nazario B1).

Academic and liberal critics charge that these emphases on abstinence-
only education and marriage as strategies for eliminating poverty obscure
important data. Biopolitical statistics produced by the Federal Interagency
Form on Child and Family Statistics demonstrate that 52 percent of U.S.
low-income working families in 2006 were headed by married couples. Yet
conservatives and neoliberals continue to insist that child poverty derives
from selfish parents who choose to procreate outside of marriage or who
choose to divorce. The gendered political economy prevailing in the United
States that compensates all but professional/managerial women at near-
poverty-level wages never surfaces in neoliberal and conservative policy discourse. Conservatives, in particular, rarely look beyond the surface data on
child poverty because of an often-unspoken organizing belief that single
mothers are inadequate disciplinarians and are thus incapable of producing
citizens who unquestioningly respect authority.

Economy—deindustrialization, automation, and globalization—plays
little to no part in neoliberal and conservative accounts of child poverty.
At the turn of the twenty-first century, "values" circulate in the popular
imagination as the first principle undergirding unemployment, underemploy-
ment, and poverty. The prioritization of values seems inexplicable given the
sheer volume of media accounts of job losses resulting from globalization
and automation. However, fragmented media reporting disconnects "social
issues" such as poverty from "economic issues" such as globalization. It is rare
that a reporter dares to connect disparate reporting categories in mainstream

press and Internet news reports. Those unusual reports that do establish linkages across social and economic issues tend to rely on values to explain linkages. In particular, reporters are apt to cite the cultural valuation of education in families as predicting individuals' economic successes or failures.

It is not surprising, then, that a diffuse but growing sense of economic unease and anxiety among white, working-class populations was addressed politically in the form of "school reform" policies and rhetorics in the 1990s and early twenty-first century. Educational reform emerged as an important strategy for guaranteeing the viability of upward lifestyle aspirations for the increasingly vulnerable "blue-collar" population and for the increasingly anxious middle-class population. Beginning in the 1990s, so-called liberal policy makers and conservatives joined forces to eliminate poverty through educational reforms. While some reforms emphasized curricular overhauls, particularly in the areas of science and math, others advocated for a renewed emphasis on school discipline and student accountability. These reforms will be explored in greater detail in the next chapter. However, what is of relevance for the purposes of this discussion is that educational reforms were viewed as a strategy that could simultaneously alleviate poverty and create jobs. An educated workforce was understood as possessing the capacity to bootstrap itself into jobs, primarily through job creation in the form of entrepreneurial initiatives. The twin objectives of school accountability and student discipline shaped federal educational policy in the form of the "No School Left Behind" program in addition to spurring statewide high-stakes testing. The emphasis on student accountability promoted disciplinary strategies and punitive measures for governing unruly or poorly prepared schoolchildren. Thus, the helping-hand ethos slowly ceded to a newly emerging punitive and disciplinary approach to child government.

Hispanic children, who emerged as rhetorically important culprits for declining educational outcomes, figured importantly in this new punitive approach. Poor Hispanic populations migrated in the 1990s and early twenty-first century to a variety of low-wage, non-unionized manufacturing jobs throughout the United States, particularly in the Southwest and Midwest. Agriculture and meat-packing plants were particularly popular destinations since local populations often rejected the poor work conditions and remuneration characteristic of these industries. However, although management in these industries welcomed these willing migrants for their hard work and reluctance to speak out against poor conditions, local communities did not. These communities often felt inundated by poorly paid workers lacking health insurance. Often, the communities lacked qualified English as a Second Language (ESL) educators. Local white groups began to demand immigration controls and reinforcement against the immigrant workers (Jordan B2). Nationally, educators began to cite the high drop-out rates of Latino students, which in 2004 approximated 30 percent. Activists and churches sympathetic to the plight of Latino workers displaced by NAFTA urged assimilation and remediation for Latino students. However, white public sentiment against immigration became increasingly confrontational and hostile, discouraging the development of federal and state programs for assisting poor immigrant populations. The rhetorical and material battles over immigrants distracted public attention from the neoliberal economic reforms and trade policies prompting the globalizing and outsourcing that together produced both job loss and south-north migration.

Criticism of NAFTA and other neoliberal economic reforms were muted until the onset of the economic recession in 2007. A strong culture of enterprise had captivated the nation's collective imagination until its material foundations began to crumble for middle-class populations in 2007. In addition to pushing school reforms, the culture of enterprise (duGay 151–69) born in the 1980s promoted market solutions for employing those pushed out of both manufacturing and welfare by stimulating private investment in declining communities. This approach rejected traditional strategies of job creation involving infrastructural improvements and entitlement programs (see Cummings 399–493).

Viewing "poor neighborhoods as underutilized markets in need of private sector investment," post-1980s government poverty-alleviation strategies worked to inject private financial capital into declining neighborhoods by identifying the competitive advantage of these areas and then creating incentives for private investment (Cummings 399, 401). Job creation was purportedly achieved through expanded local business opportunities and through support of local microenterprise projects (404). In 2006 Michael Sherraden of Washington University championed microenterprise and microinvestments in poor neighborhoods, explaining in the *Wall Street Journal* that a "small amount of wealth…enables a family to move to a better neighborhood, or enables an investment in education or training that wouldn't happen otherwise, or enables a purchase of a home" (cited in Wessel "Changing" A10).

Market-based solutions to community development also included affordable housing and lending. Affordable housing was promoted through provision of tax credits for private investment in affordable housing development and rehabilitation (Cummings 440). Financial resources were made available through the provisions of the Community Reinvestment Act (CRA) and the Community Development Financial Institutions Act (CDFI Act) of 1994, which together encourage depository institutions to provide credit to communities, including low- and moderate-income communities (http://www.federalreserve.gov/dcca/cra/).

However, Cummings argues that these localized, bottom-up approaches to poverty alleviation tend to privilege market logics to the detriment of social welfare policies and labor protections. Consequently, they have been ineffectual in addressing more macro "structural determinants of poverty" and fail to promote a "redistributive, worker-centered agenda" (454–55). Rejection of redistributive policies by public policy and market practices has exacerbated growing economic inequality. In 2005 the wealthiest 1 percent of Americans earned 21.2 percent of all income according to data from the Internal Revenue Service, while the bottom 50 percent of Americans earned 12.8 percent of all income (Ip A3). The proposed solution for this historic gap was the market-based technology of reduced taxes, although taxes fell further relative to income for the top 1 percent of earners.

Poor and near-poor children in the United States face declining social mobility and increasing insecurity in their daily lives at the beginning of the twenty-first century. Yet the helping-hand ethos has been replaced by a discourse of personal responsibility and a more punitive approach toward governing risky or dangerous children. For example, statistical representations of working-class failure to achieve often elicit calls for more personal accountability and punitive consequences. High school drop-out rates approximating 50 percent in many urban cities engender calls for the elimination of welfare subsidies for dropouts and the public shaming of "failing schools." Little public sympathy seems to exist for those who fail to conform to the demands of neopersonhood. A new biology of personhood has evolved that conveniently locates failure in bad genes, minimizing the role of environmental mediations. This reinvention and retrenchment of nineteenth-century attitudes is a primary focus of chapters 3 and 4.

CONCLUDING THOUGHTS AND DIRECTIONS

Nineteenth-century formulations of poor children emphasized dangerousness, pre-delinquency, and potential degeneracy. These formulations prompted development and institutionalization of what Donzelot called the "tutelary complex." Over time, this complex helped shift societal attitudes toward poor children, encouraging perceptions of them as victims of adverse circumstances. The early twentieth-century concern with the vitality of the population and pro-eugenic policies led to the institutionalization of social welfare programs aimed at "helping" poor children "at risk" particularly through the technologies of health (e.g., nutrition programs) and through pedagogical disciplines aimed at mothers at home and children in the schools. This chapter explored developments in the tutelary complex across the nineteenth and twentieth centuries, including child-centered Progressive reforms, the child guidance movement, social welfare reforms such the War on Poverty, and education programs for disabled children. The chapter has addressed the shifting assumptions about children and childhood that have helped foster these movements and their attendant technologies of childhood government. The chapter explained how these movements and technologies have been developed and tested in relation to economic questions of market utility. And perhaps most importantly, the chapter has explored how application of representational frames, governing technologies, and disciplinary strategies has always been shaped by class- and raced-based hierarchical relations. The following chapters develop and extend the themes characterizing late twentieth- and twenty-first-century attitudes toward the government of children also exploring their implications for the government of children abroad.

Risk, Biopolitics, and Bioeconomics

As explained in chapter 2, biopolitical issues and threats to the health and vitality of the nation were, from the eighteenth century forward, conceptualized in relation to economic formulations of national strength and capital accumulation. Western biopolitical strategies for governing children have, therefore, often been formulated in relation to the framing of economic concerns. Economic imperatives and aspirations have shaped childhood disciplines and technologies of improvement in both the home and the school historically and have governed formulations of risky and at-risk children. This chapter explores how marketplace logics and technologies of government have shaped parental attitudes and childhood pedagogies at home and at school across the last four decades. In exploring the transition to the twenty-first century, this chapter addresses how social welfare approaches to framing and regulating childhood "risks" have given ways to neoliberal approaches emphasizing individualized technologies of the self, privatized market-based solutions, and outcome-based programs stressing personal accountability and financial efficiency. These transformations in formulating and governing childhood have led middle- and upper-class parents to assume unprecedented control over their children's occupational socialization. For instance, alarmed by the discourse of the "cognitive age," characterized by a cultural preoccupation with abstract expertise, upper-middle-class citizens have become hyper-responsibilized for their children's education. Many of these parents pursue their children's education from the moment of conception forward. They often favor private education and support tax vouchers for private schools. They invest extensively in educational toys and programs (e.g., tutoring). In other words, they pursue a plethora of disciplines and technologies aimed at producing the savvy knowledge workers of the twenty-first century.

Concomitantly, contemporary neoliberal and neoconservative strategies of government have often led to more punitive approaches toward the educational management of lower-income children. Thus, while children of the lower middle and working classes are also shaped by evolving neoliberal logics and technologies, their lives have been impacted in very different ways than their upper-class peers. Neoliberal reforms in government funding have shaped public schools through new regimes of accountability and privatization. "Failing school" policies, high-stakes testing, and zero-tolerance discipline programs all illustrate neoliberal approaches to managing risky childhood subjects and spaces. The economic recession beginning in 2007 will exacerbate the challenges facing poor children as social welfare supports come under further assault.

This chapter compares the discourses, disciplines, and technologies of the self linking the market with the government of poorer and wealthier children, primarily but not exclusively through pedagogical or pedagogically inclined practices in the home and schools. The chapter begins with an account of the "cognitive age" and the attendant requirement for "knowledge" workers. It then explores how upper- and aspiring middle-class parents have become "responsiblized" for optimizing their children before turning to explore how risk-management strategies have been applied to contain the risks posed by those from the lower socioeconomic classes, who are increasingly being regarded, once again, as the "dangerous classes."

The "Cognitive Age," Neoliberal Logics of Globalization, and Families at Risk

In May 2008 David Brooks ran an op-ed column in the *New York Times* titled "The Cognitive Age," arguing that public debate about globalization misses the real significant societal change, which is the rising importance of education and intelligence:

> We're moving into a more demanding cognitive age. In order to thrive, people are compelled to become better at absorbing, processing and combining information. This is happening in localized and globalized sectors, and it would be happening even if you tore up every free trade deal ever inked.
>
> The globalization paradigm emphasizes the fact that information can now travel 15,000 miles in an instant. But the most important part of information's journey is the last few inches—the space between a person's eyes or ears and the various regions of the brain. Does the individual have the capacity to understand the information? Does he or she have the training to exploit it? Are there cultural assumptions that distort the way it is perceived?

Brooks observes that the cognitive age "emphasizes psychology, culture and pedagogy—the specific processes that foster learning." Thus, he concludes, "different societies are being stressed in similar ways by increased demands on human capital," so individual "anxiety is not being caused by a foreigner." Brooks's argument about the cognitive age is part of a larger discourse about changing market circumstances and their implications for individuals. As implied in Brooks's comments, this discourse presumes that there is a "correct" cognitive-cultural style that allows maximum exploitation of information.

While globalization is a term fraught with many meanings, the economic policies that encouraged greater financial flows and outsourcing of U.S.

economic production in the closing decades of the twentieth century were not accidental but rather were part of the larger neoliberal approach to market deregulation. As explained in chapter 2, neoliberal logics of government began to govern public policymaking in the 1980s and 1990s with welfare reform and deregulation of financial markets and trade. Neoliberal ideas can be traced to Friedrich A. Hayek's (1–283) and Milton Friedman's (1912–2006; 1–208) twentieth-century economic and political philosophies. Their attitudes toward the market had strong ideological appeal in the 1980s and 1990s, when computers and automation were transforming the organization and production of work. Automation of production and deregulation of trade and finance contributed to the erosion of labor wages and power.

Offshoring and automation together contributed to the loss of many unionized manufacturing jobs in the United States. As of 2007 only 12.1 percent of employed wage and salary workers were union members, according to the U.S. government Bureau of Labor Statistics. Of this figure, 35.9 percent were public sector workers, and only 7.5 percent were private industry workers. Comparatively, overall union membership in 1983 was at 20.1 percent (U.S. Department of Labor "Union Members"). From the 1970s onward, wages for workers engaged in routine tasks in factories and offices slowly eroded. Many routine jobs simply disappeared, although losses were not restricted to routinized work (Wessel "Why Job" A2). A 2006 survey of more than 200 multinational corporations found that 38 percent planned to "change substantially" the global distribution of their research and development projects over the next three years, potentially leading to significant job losses in the United States (Lohr). The outcome of these trends and their widespread publicity in the media heightened the public's sense of economic anxiety.

At the close of the twentieth century, public and expert sentiment about the future of work in the United States was increasingly pessimistic. A 2004 Rand Report, *The 21st Century at Work,* observes that the fifteen occupations projected to have the largest absolute increases in employment through 2010 include food service workers, customer service representatives, retail salespersons, nursing aides, orderlies, and attendants (Karoly and Panis 203). Occupations projected to grow the fastest include personal and homecare aides, medical assistants, social and human service assistants, and home health care aides (203). As the Rand authors observe, none of these occupations requires postsecondary education, and most tend to be regarded by economists as "lower skilled" and low-wage positions.

The Rand report also notes that the "labor market may be shifting toward less job stability" (Karoly and Panis 203). Future labor markets appear to be evolving toward greater reliance on small business forms such as "freelance work and self-employment" (204); thus the report suggests that governments consider providing self-employment assistance programs that promote microenterprise (204–205). The report notes that the trend toward less stability requires all workers to pursue "continuous learning throughout the working life" (205–206). The apparent contradiction between (1) trends in job growth and (2) the need for enhancing the K-12 science and technology educational curriculum and expanding opportunities for continuous lifelong learning is never fully explicated, but the report's implicit supposition is that an educated workforce has the capacity to produce its own knowledge-intensive jobs.

Public anxiety about the stagnating and declining average wages of workers in the last three decades of the twenty-first century was somewhat contained as workers worked more hours and as women entered the workplace. These trends maintained household income levels from the 1970s onward despite the decline in real wages for approximately 80 percent of all U.S. workers. Those workers who did experience wage gains tended to be professional and high-level managerial; many were employed in finance. The emerging "gold-collar worker" was often celebrated in the media thereby, encouraging the perception that education and hard work still delivered a high-consumption lifestyle (Kelley 109).

The growth of celebrity culture in the popular media seduced Americans with images of the rich and famous. The so-called reality media popularized at the turn of the twenty-first century promised everyday individuals the potential to enter into the ranks of the rich and famous. However, the vast majority of the U.S. population lived vicariously through these types of images. Cheap imported substitutes for designer goods sold at discount stores offered consumers means for demonstrating consumer power and for participating in mass-mediated lifestyles. The link between eroding wages and globalization of production of these same products escaped widespread notice or commentary despite efforts by labor activists to publicize instances of labor exploitation in poor nations.

The historically unprecedented expansion of credit also operated to mask declining wages (Wolff). Citizens' indebtedness grew substantially through credit-card-mediated consumption and home equity withdrawals and borrowing enabled by the early twenty-first-century housing bubble. The financial services industry seized upon and encouraged ever greater levels of debt. Growing public indebtedness began to raise an alarm among some economists, but this alarm was muted as household debt was recycled through mortgage refinancing and credit card transfers. Financial service institutions profited by securitizing debt, which was distributed across the global economy. Similarly, the U.S. government recycled its debt through the sale of government bonds, especially to China.

At the beginning of the twenty-first century, social critics and journalists amplified their concerns about growing social inequality using demographic statistics to paint a picture of economic stagnation for the vast majority of the population. National well-being was represented as "at risk" from this inequality. For instance, the *Washington Post* reports in 2007: "Income inequality remains at a record high. The share of income going to the 5 percent of households with the highest incomes has never been greater....(Aizenman and Lee A3). Likewise, the *New York Times* reports that "just over half of household income was concentrated in the top 20 percent of Americans in 2006, about the same as 2005. Households in the lowest 20 percent, on the other hand, accounted for only 3.4 percent of the nation's household income" (Goodnough). Economists observe that the poverty rate in 2006 was 24 percent for African Americans, 20.6 percent for Hispanics, 10.3 percent for Asians, and 8.2 percent for whites. Intergenerational data published in 2007 suggest that upward mobility for African Americans has declined (Fletcher "Middle-Class Dream" A1).

Biopolitical data collected by concerned sociologists and economists began to point to a great risk-shift that had occurred as social welfare benefits were shifted from government and employers to individuals (Hacker 1–194). One such "risk" is health insurance. U.S. adults and children increasingly lack employer- and government-sponsored health insurance; approximately 47 million Americans lacked insurance in 2006 (Goodnough). The number of children without health insurance increased to 8.7 million, or 11.7 percent, in 2006 despite establishment of the State Children's Health Insurance Program (SCHIP) in 1997. Another risk identified in this critical discourse is pensions. Many employers have stopped offering defined-benefit pension programs. For example, in 1980 approximately 40 percent of private sector jobs offered pensions compared with only 20 percent in 2005 (Lowenstein). In the early twenty-first century, financial experts warn that the security of existing public and employer-based pensions is questionable because of years of underfunding (Lowenstein; Walsh "Once Safe," "Actuaries").

Another important shift in risk concerns educational and vocational socialization. As Brooks points out in his article on the cognitive age, cited earlier, the skills required of elite workers are more diverse, dynamic, and exacting than those of the past. Professional occupational socialization typically requires advanced college degrees, and career success is often contingent upon access to prestigious universities. College-educated liberal arts majors struggle to find relevance in an increasingly lean and technocratic workplace, while those with technical degrees are required to demonstrate the social adaptability and communication competence formerly associated with liberal arts students. Exacting performance standards face all workers; workers are required to be flexible and to engage in continuous learning as their jobs are reengineered, downsized, outsourced, globalized, or otherwise reconfigured. Blue-collar workers must assume the well-publicized financial risks of failing to achieve a college degree. Many such workers manage the risks posed by deindustrialization, globalization, and automation by assuming several part-time positions in the service economy.

As explained previously, governmentality scholarship tends to avoid commentary on the "reality" of risks, focusing instead on how risk is constituted in social logics and problem-solution frames and subsequently governed through policies and programs. The risks identified in this section are typically trivialized by neoliberal economic and political authorities. Neoliberal economic policy rejects collectivist approaches, including Keynesian demand-focused economics and redistributive taxation. Neoliberals favor monetary macroeconomics that work by controlling the money supply, because neoliberals want to control inflation in order to reduce capital losses for wealth holders. Additionally, neoliberals believe "regulation" poses unacceptable risks for market expansion by stifling enterprise and innovation. Finally, neoliberal authorities favor shifting risk to individuals away from institutions because this shift in risk is seen as encouraging individual responsibility, personal initiative, and economic innovation. Redistributive policies are seen as stifling the competition believed necessary for markets to operate optimally. Pursuit and exploitation of risk are also seen as vital to the competitive operations of markets. Neoliberalism, as Foucault points out, elevates competitive market exchange as the essential mechanism for evaluating the worth of all things (*Biopolitics* 46).

Individuals at the beginning of the twenty-first century are obligated to assume responsibility for forms of risk previously collectivized. Individuals' willingness to assume responsibility for managing risks is encouraged by a proliferation of social discourses that guide, advise, and prod (Baker and Simon 1). Declining government and corporate investments in the social infrastructures of health, retirement, and education leave individuals little choice but to shoulder more risk and responsibility. Individuals who fail to choose responsibly, or whose economic failures seem to imply irresponsibility, are cast as risky subjects requiring supervision and control. In what follows, these arguments are examined in more detail in relation to the economic responsibilization of upper- and middle-class children on the one hand and the surveillance and control of lower-class children on the other.

The Cognitive Age, Aspiring Parents, and Knowledge Education

Aspiring, responsible, middle- and upper-class parents are today mobilized to ensure their children's future economic success in a risky economic environment. Parents hope, above all else, that their children will become "top thinkers" and struggle to ensure this outcome (Cohn A23). Twenty-first-century parents' mobilization can be traced genealogically to the nineteenth- century tutelary complex explained in chapter 2. In the mid-twentieth century, the "helping hand's" fusion of psychoanalysis and humanistic psychology created the psyche as a space for careful parental surveillance and engineering. With the birth of the cognitive age, the child's intellect emerged as a site for strategic engineering as well. During the 1960s- and 1970s-era War on Poverty, social welfare authorities tried to compensate for the (purportedly) intellect-dampening effects of the "culture of poverty" by drawing upon new concepts and processes explained by cognitive psychology. New compensatory programs were created (e.g., Head Start) by activists who believed that social mobility could be improved through individualized educational interventions. However, advocates' focus on the poor's "culture of poverty" eventually eroded public support for social welfare programs because this framework ultimately identifies value and lifestyle orientations as responsible for the reproduction of poverty, rather than a lack of economic opportunities. Although the research conclusions on how to engineer intellectual outcomes were never fully implemented in compensatory education programs, they were deployed in individualized technologies of the self as middle-class mothers sought to optimize their offspring's intellectual capacities.

In *The Politics of Life Itself,* Nikolas Rose describes the emergence of technologies of life that act in the present to "secure the best possible future for those who are their subjects" (6). Implied within these technologies are idealizations of optimal individual or collective life. Technologies of childhood optimization in the 1960s and 1970s focused primarily on stimulating the child's mental development through strategic social and academic pedagogies. Today, greater economic competition has produced more exacting optimizing technologies for middle-class children. Technologies that once operated from the outside upon children's psyche using the early metaphors of the cognitive age now seek to act from within upon children's brain's using the metaphors of cognitive neuroscience and neuropsychiatry. This section explores these developments by first addressing how cognitive psychology was employed to remediate poverty before turning to examine how the research framework transformed responsible parenting and mothering.

Birthing the Cognitive Age: Remediating Poverty

David Brooks was hardly the first to herald the ascendancy of the cognitive age. In 1969 Zbigniew Brzezinski (b. 1928) announced the dawn of the "technetronic society: a society that is shaped culturally, psychologically, socially and economically by the impact of technology and electronics, particularly by computers and communications" (141). Although no clear boundaries exist, the cognitive/technetronic age was born with the launch of Sputnik, the infusion of cognitive psychology into childhood pedagogies, and the growing use of computers in the workplace. These historical developments created new concerns about middle-class childhood and produced new problem-solution frames for social policy and parenting alike.

Education became a national security priority after the Soviets launched Sputnik in 1957 (Ehrenreich 1–292). Military strategists, politicians, and educators linked Cold War competitiveness to the nation's capacity to educate a technically and scientifically skilled workforce. In 1958 the U.S. Congress passed the National Defense Education Act (NDEA) (PL 85–864) aimed at securing the nation through "the fullest development of the mental resources and technical skills of its young men and women" (cited in Flattau et al. ES-1). Educators were mobilized by government grants and public discourse to enhance the nation's overall educational standing. Risk was in a sense tied to national educational standing, and, therefore, social welfare policies could be legitimized within economic calculi of value.

Nikolas Rose argues in *Governing the Soul: The Shaping of the Private Self* that experts' answers to the problem of engineering intellectual excellence were shaped by the social welfare problem-solution frame of "compensatory education" aimed at breaking "the cycle of poverty" in the context of the 1960s- and 1970s-era War on Poverty (194). The child's preschool years increasingly came into focus as a "critical" temporal site for the development of intellectual and linguistic proficiencies. For poor children, preschool would emerge as an important site for remediation; in contrast, middle-class parents adopted preschool as a strategy for optimizing their children's intellectual competitiveness. This contrast between remediation and optimization characterizes the government of working-class and middle-class children across the last four decades of the twentieth century. Cognitive psychology was instrumentalized for both remediation and optimization.

Cognitive psychology explained how experience shapes thoughts and problem solving. Jean Piaget's (1896–1980) theories of children's cognitive development played an important role in popularizing the cognitive approach. Piaget's framework blended experiential knowledge (i.e., sensory experience) with the child's (believed) a priori mental organizing capacities to describe transformations in cognition across developmental stages (see *Language* 1–276). Piaget's influence on American cognitive psychology was particularly evident in the developmental research on language acquisition, perception, and intellectual processing conducted in the 1950s and 1960s (Kessen 168–69). Some research on these themes held that cognitive skills could be improved by intensive engineering of social environments, since environments shape cognitive "inputs" and thereby impact processing categories. For instance, mothers' language skills, particularly their use of abstract language, were seen as having a significant effect on very young children's cognitive development by fostering abstract categories of thought (e.g. Olim 53–68). Cognitive psychology therefore provided social welfare advocates with a tool kit for grappling with the challenge of jump-starting the disadvantaged child in the absence of broad socioeconomic reforms.

Educational interventions emerged as the most important approach for breaking the cycle of poverty. Specially devised pedagogies were developed to overcome culturally produced deficiencies believed to hamper the poor's mobility (Blank and Solomon 47). Educators' faith in the role of remedial pedagogy prompted structured experiments aimed at demonstrating gains in measured IQ scores, as illustrated by Merle Karnes and Audrey Hodgins's 1969 study, "The Effects of a Highly Structured Preschool Program on the Measured Intelligence of Culturally Disadvantaged Four-Year-Old Children" (89–91). Although not all of the early research supported a link between "enriched" environments and later cognitive complexity, the public and academic imaginations were captivated by the posited relationship (see Wachs and Cucinotta 542). Educators advocated for more funding for educational interventions targeting "early critical years of cognitive development" to compensate for the purported deficiencies of poor mothering (Olim 53).

This belief that educational interventions alone could triumph over poverty implicitly suggested that the poor suffered from an intellect- and initiative-dampening environment. On the one hand, environmental explanations of intellectual and academic achievement were viewed as progressive because they countered racist, hereditary accounts. However, on the other hand, most popularized environmental accounts ended up blaming the poor by attributing their poverty to a deficient and sometimes degenerate culture of poverty. This tendency to stigmatize the poor by ascribing their poverty to deficiencies in their *cultural* environment is evident even in relatively progressive efforts to combat racism.

For instance, Ashley Montagu's article "Sociogenic Brain Damage" published in 1972 illustrates these ambivalences in the era's environmental accounts. Montagu's essay begins by contesting Arthur Jensen's genetic approach to explaining intelligence. In 1969 Jensen had published an essay arguing that genetic factors strongly explain measured differences in the IQ scores and scholastic achievement of African Americans and whites (82). Jensen suggests in this essay that compensatory education fails to overcome genetic limitations. In contrast, Montagu asserts that "the damage capable of being done by unfavorable socio-economic conditions to the development of intelligence is even more extensive and conclusive" than hereditary contributions (1054). Montagu's environmental framework stresses physiological deprivation by describing how biological inputs such as nutrition shape neurological development (1045–1047, 1050). However, Montagu also describes a "social deprivation syndrome" characterized by "short attention span and learning difficulties resulting in poor test performance" (1052). He observes that these "deficiencies" are often inaccurately explained by hereditary factors when in fact they are produced by social deprivations during critical developmental periods. In this fashion, Montague implicitly condemns the poor's parenting skills as "deficient." Moreover, Montague takes a rather pessimistic view of poverty's neurological effects, arguing that brain damage can be "more or less irreversible" (1047). In sum, Montague's model of environmental deficiencies implicitly pathologizes the poor's parenting but holds out the possibility of environmental interventions during early critical periods.

Montagu and others social welfare reformers extended early child savers' mantra of prevention using new vocabularies and new problem-solution frames. They believed that poverty could be eliminated, or its effects mitigated, by engineering children's environments using newly developed cognitive tools. Structural economic reforms were not necessary to accomplish environmental interventions; rather, change could be accomplished at the level of the individual child through specific expert interventions. Hamilton Cravens observes that intervention efforts during this period in the 1960s and 1970s were based on a novel, individualistic conceptualization of the child (21). Only by individualizing the child in terms of perceptions of risks and opportunities could advocates conceivably argue for the defensibility of specific interventions. Cravens illustrates this argument using Head Start as an example:

> Head Start clearly was premised on assumptions inconceivable in the period between the world wars, that the individual, if caught early enough in the life cycle, could be trained to "jump" from the norm of the group or subculture to which he or she was born, to that of an entirely different group. (21)

Targeted, specific interventions were thus believed capable of compensating for the global deleterious effects of poverty on child development.

Cognitive psychology thus fit neatly into the aims and technologies of a century-old tutelary complex that elevated education as the most important strategy for governing the poor. The project of compensatory education was in many ways progressive because it combated deterministic hereditary accounts of poverty. Yet, compensatory education was seen as effective only during early critical periods, rendering older poor children beyond the scope of remediation. Additionally, the project of compensatory education often implicitly invoked and promoted a "culture of poverty" approach that largely ignored economic factors (except nutrition) and therefore pathologized the poor's knowledge bases, values, and lifestyles. This "culture of poverty" thesis would eventually erode public willingness to expand supports for the poor in the 1980s. In effect, the impact of this project of engineering the intellect was probably greatest for middle-class families.

Middle-Class Cognitive Optimization

In the 1960s and 1970s, aspiring, middle-class parents were increasingly concerned that their children would not be prepared for the demands of the technetronic society described by Brzezinski and other such authorities (Ehrenreich 1–292). A college degree had become a prerequisite for a guaranteed middle-class livelihood, and so parents feared that their children would not gain entrance into the more competitive universities. Children had to be cultivated from their earliest years. Thus, fears about class mobility brought quality preschool into focus as an important strategy for ensuring the later academic success necessary for gaining access to prestigious universities.

Middle-class mothers sought advice on how to stimulate their children's cognitive development at home. Like their mothers in the 1940s and 1950s, the mothers of the 1960s and the 1970s turned to the experts.[1] Both generations were influenced by the "helping hand" model of child government discussed in chapter 2. The helping-hand model relies on an invisible pedagogy emphasizing the role of social-environmental supports in fostering children's development (Cox 191). While parents of the 1940s and 1950s were primarily concerned that their children conform to age-appropriate socioemotional norms, the parents of the 1960s and the 1970s wished their children to exceed normative standards and goals. These anxious parents looked to the experts for guidance on how to structure their children's environment so as to *maximize* their children's capacities (see Stearns 1–251).

Cognitive psychologists responded to middle-class angst by describing and prescribing the conditions for "optimum learning and optimal environments" (Goulet 13). Efforts to "optimize learning and performance" (Goulet 13) in typical children represented a departure from early efforts to *normalize* deviance or even to prevent the risk of future deviance in the "at risk" child. Cognitive psychology explained children's sensory perception, category formation, and knowledge processing in ways that encouraged parental surveillance and intervention. Middle-class parents were deluged with solicitous advice on how to engineer excellence by stimulating cognitive development by talking to their children, facilitating new experiences, and so on. Preschool emerged as critical time-space for middle-class children to acquire the knowledge and experiences that would facilitate cognitive complexity, not simply school readiness.

Efforts to optimize cognitive development were not restricted to preschool children but increasingly targeted infants as well. By the mid-1960s, mothers were advised to stimulate their infant's development by expanding their sensory environments. My own mother was frustrated by my oral consumption of the magazine images she had pasted inside my crib in 1965. Her efforts to stimulate my cognitive development were part of a larger movement aimed at understanding and governing the infant brain. As one text on infant development observes in 1971, "Infant research is exciting, informative, challenging, and exacting. It is still *new*, however, and its research tools are tentative" (my italics, Caplan 13). In 1972 *Scientific American* published an article by Jerome Kagan titled "Do Infants Think?" (74–82). This article brought to popular attention cognitive science's newfound interest in infant perception, cognition, and development. Kagan reports in the article that by the end of the first year, typical infants have developed cognitive structures allowing for the transformation of discrepant events into familiar forms (74). By emphasizing the acquisition of cognitive structures during infancy, cognitive psychology drew attention to the role of deprived or enriched "neonatal experiences upon later cognitive functioning" (Wachs and Cucinotta 542). In 1977 the *Annual Review of Psychology* ran an article titled "Human Infancy," in which the authors claim that the field of study dedicated to human infancy had "come into its own," as evidenced by allocation of a whole chapter to the subject (Haith and Campos 251–93).

The popular media played an important role in widely disseminating findings about infants' cognitive development. Dissemination of research studies documenting infant cognition and learning stimulated child-welfare advocates. Educators and other child advocates demanded ever earlier intervention for disadvantaged infants. For example, in 1988 the *New York Times* ran an article titled, "When Head Start Is Too Late," arguing:

> Americans are beginning to acknowledge the importance of prenatal and neonatal health care for poor children at risk. Each dollar spent on prenatal care saves three dollars in the cost of care for babies with low birth weight.
>
> The country also is learning that disadvantaged 3- and 4-year-olds who participate in Head Start and other quality preschool programs are much more likely to finish high school, hold a job, escape welfare and avoid crime. But what is not yet obvious is that for many disadvantaged children, intervention even by age 3 is already late. If these children are to succeed later in life, they need help earlier. (A26)

The article concludes:

> Public school used to be the point where society took some formal responsibil-
> ity for children. Starting in 1965 the point shifted to preschool programs, as
> society recognized the need to give disadvantaged children a Head Start on
> public education. Now, to give such children a fair chance, the point is shifting
> again, to when learning begins: birth. (A26)

Although the idea of disadvantaged infants captivated public attention, it
was the infants of the affluent who became subject to the social project of
engineering infancy.

The cultural preoccupation with infant achievement no doubt stemmed
from increasing economic anxiety. During the 1970s and 1980s, the U.S.
workplace was transformed by the introduction of computing and globalization. Simultaneously, U.S. factories were automating operations with
the introduction of robotics, as illustrated by this article that ran in the
Washington Post, "High-Tech Revolution in Robotics Under Way; Unlimited
Uses Could Transform Way People Work" (Harden H1). By the 1980s a number of reports and popular management books had publicized the increasingly exacting demands of the newly automated, computerized workplace.
In 1986 Paul Adler published an article detailing "new technologies, new
skills" required in the newly automating workplace (9–28) followed closely
by Shoshana Zuboff's 1988 account of a polarizing workplace characterized
by highly routinized work and highly demanding work. Middle-class and
upper-class parents aspired for their children to be intellectually agile, tech-
nologically adapt, socially competent "knowledge workers" hired for their
"problem-solving abilities, creativity, talent, and intelligence" necessary for
executing "nonrepetitive and complex" work responsibilities (Kelley 109).

By 1983 the pressure to raise superior children was so pervasive that
Newsweek ran an article titled "Bringing up Superbaby" (Langway et al.
62). In the article, one mother comments upon why she is willing to spend
$2000 for preschool tuition: "There's so much pressure to get into col-
lege," says 38-year-old Linda, "You have to start them young and push them
on toward their goal. They have to be aware of everything—the alphabet,
numbers, reading. I want to fill these little sponges as much as possible."
As Fitch observes in a marketing essay published in 1985: "There is a tremendous pressure for the children to have some competitive edge" (35).
Facilitating informed parental surveillance was a new version of the baby
book that helped parents track and stimulate their infant's acquisition of
physical, social, and cognitive skills. Books such as Eisenberg, Murkoff, and
Hathaway's (1989) *What to Expect the First Year* and Eisenberg's *What to
Expect the Toddler Years* (1996) advise parents to consult with their pedia-
trician if their infants fails to meet developmental guidelines outlined on a
monthly basis. Parents, formerly enjoined to be responsible for engineering
their child's intellectual and emotional development, became sensitive to any
"delays" in their *infant's* cognitive and social development.

How can we tell that economic anxiety drove this new cultural concern with infancy? One indicator is found in the metaphors used to describe infant and child cognitive development. Beginning in the 1970s, computer metaphors increasingly dominate representations of children's and infant's cognitive processes. For instance, as early as 1974, an article appeared in *PsycCritiques* titled "The Transistorized Child," which reviewed a book covering information-processing approaches to cognition in children (Cotton and Ellis 37–38). Cognition in these accounts is constituted as a "computational" process in which "mental processes are computations on formal, syntactical symbols" (Jones and Elcock 218). In the 1970s and 1980s, cognitive computation was represented as occurring in autonomous, functionally intact cognitive modules (Jones and Elcock 218). In the late 1980s, more distributed models of artificial intelligence would begin to edge out the CPU-based representations, as illustrated by Goldman-Rakic's 1987 article, "Development of Cortical Circuitry and Cognitive Function" (601–22).

Use of more distributed models of artificial intelligence to describe infant and child information processing reflects the growing popularization of brain research. Early studies on brain development primarily used animal models to test how environmental stimulation affected cortex gains and brain enzymes, which were analyzed through autopsies. However, beginning in the 1960s, new brain-imaging devices were being developed and applied to "reveal" the interiority of the brain, encouraging new, more networked views of neural activity. Experimental psychologists began to use electrocardiograms (EKG), electro-oculography (EOG), and electroencephalography (EEG) to study infant brain activity coinciding with behaviors such as eye movement and sucking (Haith and Campos 253). Eventually, these efforts were extended by attempts to link learning, attention, and discrimination with neurophysiologic measures.

By 1978 computerized axial tomography (CT), introduced publicly in 1972, was being used to produce images of right-left brain asymmetries (Galaburda, LeMay, Kemper, and Geschwind 853). CT uses computers to combine X-ray images shot from diverse angles into a synthetic image. CT was a more helpful technology for medical authorities searching for tumors and other morphological brain abnormalities than it was for cognitive psychologists looking for ways to link behaviors and traits to brain centers or circuits, but this technology was seen as just the beginning of, as one article title quips, "The Bright New World of Brain and Body Scans" (Arehart-Treichel 170–72). New scanning technologies had many applications beyond revealing the interiority of young children's brains, but it was precisely this area of research that would captivate popular attention.

BEAM, or brain electrical activity mapping, was one of the new imaging/representing technologies that tantalized cognitive psychology and attracted popular interest because of the realism of the images. BEAM was developed as a way of representing and interpreting the vast amounts of data produced through the EEG imaging technology, which measures electrical activity in the brain. BEAM condenses, summarizes, and displays on a color monitor "spectral, spatial, and temporal information on brain activity" from multiple scalp-monitoring locations and can statistically compare data with a control group (Torello and Duffy 96). In 1985 BEAM was heralded as "not only a tool to study the brain," since it "may provide new information about learning processes and supplements traditional methods of assessment such as IQ tests, neuropsychological test batteries, and clinical observations" (97). BEAM, in concert with other new technologies—such as PET (positron emission tomography) and MRI (magnetic resonance imaging)—allow for the "discovery of converging lines of evidence for the diagnosis of cognitive dysfunctions" (97).

BEAM, PET, and MRI assumed the status of truth-telling devices for the reality of the brain. Neurological imaging technologies lent scientific authority to developmental psychology's representations of children's cognitive development. Technologies such as BEAM and fMRI (functional magnetic resonance imaging) created a neurological spatial locale that was visible, measurable, and subject to intervention. Expert authorities suggested that the brain could be measured and compared across states, allowing for experimental interventions aimed at maximizing neurological activity or growth.

Educational authorities were among the first to be captivated by cognitive neuroscience. The 1978 National Society for the Study of Education yearbook was dedicated to the topic of education and the brain (Peterson 75). Four years later, the American Association for the Advancement of Science's annual meeting featured a symposium entitled "The Brain Science of Education" (75). Public educators were tantalized by the possibility of using brain science to develop pedagogical methods (see Bruer "Brain Science" 14–18, "In Search" 649–57). Educators rejected the more deterministic representation of fixed circuitry in favor of the neurological plasticity:

> Fixed Circuitry—To begin, a number of neuroscientists now have found evidence to support the notion that the human brain has a set of pathways or fixed circuitry patterns which seem to be present prenatally...But before one concludes that everything is fixed, one should know of other new discoveries which disclose the plastic or dynamic and changing nature of the brain. Plasticity—Although many general areas or features of the brain are observable at birth, neuroscientists also have discovered that some parts of the brain function differently at birth or in the first few years than in adulthood. From many sources of evidence, we know that the brain changes with age. This plastic quality is of special interest to educators as well as to neuroscientists and cognitive scientists. (Peterson 75)

Critical periods, a concept long in currency in the developmental literature, was seen as bridging the relationship between fixed circuitry and neural plasticity. Brain science explained the infant brain as plastic and flexible but cautioned that neurological rigidity set in at the ripe age of around three years.

Neuroscience's findings on critical periods implied a two-dimensional developmental model, fixing a person's neurological growth into two fixed states (Nadesan "Engineering" 401–32). The first state is early childhood (birth through age two or three), which is ordered into a homogeneous, chronological time line complete with developmental scales. As a spatialized zone fertile with possibility, this neurologically plastic first state of early childhood must be scrutinized with the intent of identifying phenomena that contribute to outcomes measured in the second state, adulthood. The second and "dependent" state—adulthood—is an indeterminate age represented asneurologically rigid, as evidenced by purportedly fixed IQ scores, or other "objective" measures of ability and professional "success."

By the 1990s the findings derived from brain scans circulated in the popular media. The infant had been transformed into the "scientist in the crib," and the revelation of the interiority of this space was the province of "cognitive science" (Gopnik, Meltzoff, and Kuhl vii). Media representations of these research findings drew explicitly on brain science to educate parents about the possibilities of optimization and the dangers of early infant "deprivation." For example, in a 1996 article *Newsweek* magazine claimed:

> It is the experiences of childhood, determining which neurons are used, that wire the circuits of the brain as surely as a programmer at a keyboard reconfigures the circuits in a computer. Which keys are typed—which experiences a child has—determines whether the child grows up to be intelligent or dull, fearful or self-assured, articulate or tongue-tied. (Begley "Your Child's" 56)

The computer analogy reinforces the implied immutability of early brain formation, particularly through the idea of neural wiring. In contrast to mental rigidity after age three, the *infant's* brain is represented as "neuralplastic" and subject to engineering, as illustrated in *Newsweek*'s article, "How to Build a Baby's Brain" (Begley 30).

In effect, brain science persuaded parents that every child could be a superchild if exposed to the "correct" stimulation at the proper developmental moment. Parents scrutinized every stage of their child's development and provided stage-appropriate stimulation designed to optimize cognitive development. By embracing these technologies, parents implicitly assumed responsibility for engineering their children's fate in the second stage of the model. Failure to achieve or exceed appropriate developmental milestones required immediate action and therapeutic remediation.

Brain science allowed advocates to argue for the market utility of public expenditures on child care in a period when neoliberal economic policies decried government spending. This usage is illustrated in a passage from the Rand Corporation's policy paper on government support of child care:

> Scientific discoveries over the past two decades have transformed the way in which researchers, policymakers, and the public think about early childhood. For example, recent research on brain science has provided a biological basis for prevailing theories about early child development, and cost-benefit analysis has reoriented some of the discussion about early childhood toward prevention programs. (Karoly, Ghosh-Dastidar, Zellman, Perlman, and Fernyhough iii)

The report includes a table of child outcome factors such as reduced child maltreatment, increased labor force participation, improved pregnancy outcomes, and monetary benefits (or costs) to government, including "lower costs to child welfare system," "increased tax revenue," and "lower costs for criminal justice system" (13). The report's decision-making rule "requires that programs or strategies produce enough savings or pay back their costs in the long run" (14). Brain science thus achieves value within neoliberal calculi of cost-saving and cost-benefit analyses.

Although the Rand Corporation promotes targeted and cost-efficient government support for specific, at-risk populations, neoliberal policies pursued by the federal government at the beginning of the twenty-first century diminished public expenditures for nearly all forms of social welfare for children. Indeed, neoliberals and neoconservatives actively combat government subsidies for child care, as illustrated by a report published by the conservative American Enterprise Institute, which argues that subsidies are market distorting (Besharov, Morrow, and Myers). In the early twenty-first century neoliberal and conservative imagination, children are cast as the ultimate responsibility of parents who are encouraged to follow the advice of educational and academic authorities regarding effective parenting in order to produce children who will eventually become economically self-sufficient adults. Many parents attempting to adopt expert advice struggle with the competing demands of expanding work hours on the one hand and intensive parenting on the other.

Risk in the Age of Optimization

As public policy makers over the last ten years gauged the cost savings of providing services to struggling working-class parents, upper-middle-class mothers struggled psychically amid material privilege to achieve the ideals of the increasingly demanding occupation of twenty-first-century motherhood. In a 2005 essay, "Mommy Madness," Judith Warner describes the "perfect madness" that accompanied her efforts to be the "good mommy":

> Back in the days when I was a Good Mommy, I tried to do everything right. I breast-fed and co-slept, and responded to each and every cry with anxious alacrity. I awoke with my daughter at 6:30 a.m. and, eschewing TV, curled up on the couch with a stack of books that I could recite in my sleep. I did this, in fact, many times, jerking myself back awake as the clock rounded 6:45 and the words of Curious George started to merge with my dreams. (92)

Likewise, Anna Quindlen describes how modern motherhood was transformed into a profession:

> Modern motherhood was codified as a profession. Professionalized for women who didn't work outside the home: if they were giving up such great opportunities, then the tending of kids needed to be made into an all-encompassing job. Professionalized for women who had paying jobs out in the world: to

show that their work was not bad for their kids, they had to take child rearing as seriously as dealmaking. (50)

Affluent mothers increasingly bear the responsibility of demonstrating their professionalized commitment through intensive mothering during infancy and toddlerhood and, later, through overt participation in parent-teacher organizations, after-school athletics, social activities, and so on.

In a time of heightened preoccupation with infant cognition, affluent working mothers had new reason to be concerned about day care. From the early 1970s forward, the media had bombarded women with messages about the potential emotional risks posed to infants and small children in child care. But brain science's findings on critical periods and infant plasticity raised new concerns about child care's effects, as illustrated by an article in the *Nation* titled "Childcare Brain Drain?" which suggests that many infants are deprived of vital stimulation while in child care:

> In an environment rich in all sorts of learning experiences the growth of synapses—the connections between nerve cells in the brain that relay information—is more lush, and this complex circuitry enlarges brain capacity. Infants who are not held and touched, whose playfulness and curiosity are not encouraged, form fewer of these critical connections. Indeed, recent research discussed at the White House conference indicates that the amount of time caregivers spend talking to infants and their sensitivity to the babies' reactions are powerful predictors of language facility and intellectual capacity. (Barnet and Barnet 6)

Given this purported link between quantity and characteristics of caregiver involvement and subsequent intellectual prowess, *Time* magazine worries that very young children in child care are at risk for "not receiving the kind of attention that promotes healthy brain development" (Collins 60).

The child care industry targeting aspiring parents responded with new advertising campaigns celebrating their programs' efficacy in stimulating cognitive development. Tutor Time became a leading child-care provider in some areas of the country, and high-end child-care providers were able to able to demand over $1200 a month for infant care by 2001 in some urban areas. Affluent working mothers concerned about their children's intellectual development paid high child-care fees, while working-class parents were forced to take the care available at fees they could afford.

Still, the perceived import of infancy led some new mothers to leave work in the 1990s, even when such a decision caused financial strain. Despite some employers' claims, "family friendly" workplaces remained far from the norm. The intensive project of mothering intellectually agile and socially adroit infants was rendered a difficult balancing act for the working mother. New social advocacy groups organized around working mothers rose in response. For example, a new organizing forum, MothersRising.org, produced "The Motherhood Manifesto," a documentary that articulates the "challenges" posted by working mothers (Jesella). The group appeals to liberal rights when demanding more workplace accommodation and nondiscrimination policies for working mothers, but also seeks alliances with organizations such as the Christian Coalition that affirm the sanctity of motherhood.

Upwardly aspiring mothers confronted a new challenge in the late 1990s when scientific authorities linked breastfeeding with children's measured intelligence scores. Infant child care not only deprived the infant of necessary intellectual stimulation, but it interfered with the infant's access to breast milk, nature's own IQ booster. For example, a *Newsweek* article, "Rooting for Intelligence," links the scientifically established nutritive benefits of breast milk to increased IQ. The claim is couched in clinical terms:

> And each ingredient has a purpose. Specific fatty acids found in breast milk have been shown to be critical for neurological development. Certain amino acids are a central component for the development of the retina, which could account for breast-fed babies' increased visual acuity—another way of measuring advanced brain development. (Glick 32)

The nature of breast milk was cast as an object of scientific knowledge, the process of breastfeeding instrumentalized as a medium of cultural capital through its capacity to increase intelligence (Nadesan and Sotirin 217–32). The science of breastfeeding offers an investment in future capital (quite literally, in the value-adding capacity of future knowledge workers).

Breast milk, as a disembodied commodity, and breastfeeding, as an instrumentalized medium of cultural capital, entered into the public circulation of goods and services as economic and political issues (Nadesan and Sotirin 217–32). The social policy journal *Governing* advises, "Breastfeeding also makes good economic sense for states interested in keeping a lid on health care and welfare costs. Florida, for example, requires an emphasis on breastfeeding in nutrition programs for welfare recipients" (Mahtesian 54). Professional human resource journals likewise advised that breastfeeding reduces absenteeism and health-care costs (Gibson). Breastfeeding is governed by policy calculations at the juncture of capitalism, bureaucracy, science, and expert authority, as women are instructed to consult their pediatricians before breastfeeding.

Breastfeeding was but one technology among many offered to aspiring parents hoping to optimize their children's potential. Parents were advised to ensure their children's adequate nutrition. Parents were advised to allow children mobility outside of infant carriers so that they could explore and discover. Parents were advised over and over to read to their children. Indeed, reading evolved as a technology of the self whose enactment would both optimize middle-class children and counter the deprivation of working-class children. Take, for example, this passage from *Time* magazine advocating reading as a strategy for redressing poverty:

> As developmental experts often point out, child rearing is not an innate skill, and several states are trying to help educate parents about parenting. Home

visits by social workers or nurses are among the most promising methods. In Oregon such visits occur under a program called Healthy Start. Sandra Daus, 22, a single mother of an 18-month-old girl, recalls the help she received from Melissa Magill. "She encouraged me to read books, a lot of books," says Daus. "I thought when Sydney got older, maybe two or three, we'd start reading. Melissa said no, start reading to her now. Sydney was a month old." (Collins 61)

In *Time's* account, reading alone counters socioeconomic deprivation by ensuring poor children's educational "readiness" and eventual capacity for economic contribution. In effect, reading at one month is heralded as the means to eliminate barriers to class advancement. Therapeutic programs directed toward rehabilitating working-class parents by teaching them to read to their infants are held to erase the many structural barriers posed by poverty and marginalization.

Perhaps the most popular approach for stimulating children's intellectual development in the last two decades of the twentieth century involved cognitive-development toys. Toys' appeal to parents as devices helpful for training children is hardly new. Early twentieth-century toys such as the Erector set marketed in 1913 began to link toys to specific occupations such as engineering. Chemistry sets appealed to parents who wished their sons to become scientists in the 1950s and early 1960s. By the mid-1960s toy ads had become very explicit in promising to develop technical skills required for the increasingly technical workplace. For example, ad copy from a 1964 catalog promises to give children leverage in the competitive educational world through toys "designed by an architect with mathematical and geometric learning implications" (Cross 160). As Cross puts it, "The educational toy was part of a great progressive project to transform society into a new order ruled by rationality and appropriate tools" (136). Educational toys assumed even more importance at the turn of the twenty-first century in the context of heightened anxiety about intellectual optimization.

Very young children need to develop abstract cognitive dexterity as well as practical programming skills. Toy advertisements and packaging promise these capabilities. An article titled "Handing Baby a Start on Genius" illustrates the role of consumable goods in stimulating development:

Psychologists at Duke University say they have discovered that if you slip a pair of Velcro-covered mittens on an infant's hands, and then put some more Velcro on toys, the child can pick up the toys, instead of just batting them. After a while, the psychologists say, the child begins to explore objects in a more sophisticated manner than do babies who are sticky-mittenless…the news release announcing the study…described the mittens as giving infants "a developmental jump-start." (Nagourney A12)

New marketing campaigns were also launched around infant developmental media, such as the Baby Einstein line of videos, and around infant and toddler computer programs and computerized toys. Novel toys were marketed to parents and children on special child-friendly television channels such as Nickelodeon and Disney.

Nelson-Rowe contends that parents' expressed toy preferences parallel two marketing strategies for educational toys. First, one category of educational toys is defined through its promotion of general skills, traits and values including, according to Nelson-Rowe, "hand-eye coordination, fine-motor skills, and concentration, as well as such general traits as creativity, imagination, and discovery" (119). For preschool and older children, this first category of educational toys also includes toys that purport to foster "social skills such as sharing and cooperation" (119). The second category of educational toys is defined by its claims to foster "specific intellectual skills," which include "reading, writing, and mathematics, as well as knowledge of colors, time, and science" (119). Also included in this second category would be computer programs such as ReaderRabbit Toddler, Jumpstart Baby, and Jumpstart Toddler, which promise parents to teach computer skills such as "mouse control" in addition to colors, shapes, numbers, and letters. The president of Vtech Industries (which produces electronic toys), Rick Mazursky, argues that "parents want their children to excel, and they know they have to be involved in technology in order to do that" (cited in Gubernick and Matzer 82). Children's receptivity to computers delights parents until fears about overuse set in, as parents confront the perils of computer "addiction" (St. George "Study Finds Some Youths" A2).

Television consumption also produces middle-class parental ambivalence. On the one hand, "educational" programs such as *Sesame Street, TeleTubbies, Blues Clues, Dora the Explorer,* and *Barney* (broadcasted on Nickelodeon, Disney, and PBS channels) promise intellectual stimulation through television consumption and through play activities with a wide range of branded toys. "Christians" have even created their own educational television programming with *VeggieTales* and *McGee and Me,* which promise to protect childhood innocence while also promoting cognitive and moral education. On the other hand, overconsumption of television—even child-friendly television—is perceived as adversely impacting young children's imagination, personal creativity, and outdoor play.

Television for young children deflects adult criticism by promising to teach educational preparedness. In contrast, television for kids over the age of six makes no such promises. Instead, this programming appeals directly to kids. For instance, Nickelodeon's edgy shows often involve secretive childhood worlds hidden from adult characters, as illustrated in *Jimmy Neutron, Fairly Odd Parents,* and *BenTen* (Banet-Weiser 215). Parents object to older kids' programming because of rebellious themes and subtly sexualized images, which together threaten childhood innocence. Cultural critics observe that children's programming encourages children to define themselves in terms of mediated identities and lifestyles that simultaneously encourage liberal values/lifestyles (i.e., consumption choices, freedom, individualism) while also invoking traditional and religious-inflected gender roles and Manichean themes of good versus evil (see Hendershot 1–261). The occasional resistant show or episode such as *SpongeBob Squarepants* (192–208) is typically overwhelmed by traditional images of combative masculinity and decorative femininity. Despite consternation from both conservative and liberal social critics, little effort is made to enhance educational programming for older kids. Only the youngest children require educational programming.

The contemporary concern about young children's television exposure can be traced to late twentieth-century anxieties about cognitive development. By the 1980s, educators and upwardly mobile parents were convinced that adult IQ scores hinged on children's early educational experiences. Careful social engineering was called for in order to maximize cognitive developmental outcomes.. Agencies such as New York's IvyWise Kids advise aspiring parents on optimal preschool placement. Affluent parents compete for slots at desirable preschools and hold their children back from starting kindergarten in order to give them an edge over their peers. Some parents even bribe school authorities to ensure their child's access to highly coveted schools. The perceived need for educational excellence is so pressing that parents will even relocate their families "in search of a perfect private education" (Hwang A1). Accordingly, one parent explains the rationale for uprooting his family: "To me, it's better than leaving them a house or my 401k" (cited in Hwang A1).

Middle-class confidence in public schools declined as anxiety about children's intellectual performance increased. "Gallup's Pulse of Democracy" reports in 2008 that almost 20 percent of U.S. children either attend private or parochial schools or are homeschooled. Private-school vouchers grew in importance as a salient political issue not only among religious conservatives but also among upper-middle-class Americans fearful of what they perceived as declining public schools.[2] In the *Two Income Trap,* Elizabeth Warren and Amelia Warren Tyagi claim that parents' concern about school quality drives housing decisions, leading families to leverage themselves financially when purchasing homes (25). By the early twenty-first century, public schools in all but the best areas are perceived as fraught with academic and security risks.

Efforts to improve children's education and to compensate for the deficiencies in their educational experiences lead many middle- and upper-class parents to schedule their kids' after-school time. Middle-class kids today are enrolled in after-school sports, tutoring, music, and a wide array of other activities. Some social critics respond that this type of harried lifestyle put children "at risk" of stress through overscheduling (e.g., see Rosenfeld's *The Overscheduled Child*). An entire line of research was launched exploring the risks of upper-middle-class childhood activities. One study concludes, "Kids' activities linked to success, not stress," but cautions that "parents" might be "overloaded" by the "struggle to balance kids' commitments" (St. George "Study Finds Kids" A21).

Not surprisingly, given the range of extracurricular activities and the consumption of educational toys, middle-and upper-class children in the U.S. are recipients of unprecedented spending. Department of Agriculture estimates in 2008 for total spending to raise a child through age seventeen were $298,000 for high-income families, $204,000 for middle-income families, and $148,320 for low-income families ("Cash-Hungry Kids 13). New concerns are raised that these privileged children, who received so much in the way of material goods, lack financial discipline. Articles began appearing in women's magazines and daily newspapers advising parents on how to raise financially savvy kids, as illustrated by this article: "Helping Kids Balance the Facts: Starting a Financial Education Early is Key to Raising Money-Wise Children" complete with a "Hands on Banking program by Wells Fargo" (Moravcik B4).

Randy Martin's *Financialization of Everyday Life* argues that aspiring middle-class parents conclude that their tendency to shower children with consumer objects undermines the kids' capacity for deferred gratification and their future financial savvy. As Martin explains, parents conclude that the "American dream of abundance should not be inhabited by children, but should remain a dream deferred" (157). Parents fearful of producing imprudent consumers turned to a range of "new pedagogies of home economics" in the form of software programs, allowance protocols, and educational literature aimed at producing financially savvy marketplace actors. This pedagogy of financialization resonates with parents faced with the demands, ambiguities, and post-2007 horrors of managing their own retirements as their employers shed defined-benefit pensions in favor of employee-managed 401(k)s.

Middle-class and affluent parents have recently confronted another risk to their offspring's future success: childhood optimization efforts produce a sense of self-entitlement among teenagers and young adults (Twenge B5; "Today's Youths" A15). College professors frustrated with the entitlement attitude of the Millennials, also called Generation Y, have conducted research demonstrating an "epidemic of narcissism" (Twenge B5). Students accustomed to near-continuous adult reinforcement and praise apparently lack the capacity to exercise the type of personal self-discipline required for mastering academics without special accommodations and guidance. The main problem with these students, according to Jean Twenge, is that "many will find they lack the work ethic and attitude necessary to succeed" (B5). Concerns about the entitlement ethos of America's middle-class youths, first outlined by Christopher Lasch's *Culture of Narcissism*, achieve new salience as privileged children apparently lack the personal fortitude and work ethic necessary for competing successfully in the Darwinian marketplace.

Optimizing the At-Risk Affluent Child

Middle-class and upper-class infants and children who fail to meet standardized norms of behavior, intellect, or affect are targeted for expert intervention. A trip to the pediatric office often ends in a referral to a child psychiatrist or psychologist for further evaluation or medication. Even infants are increasingly targeted for psychiatric intervention (including drugs) to "head off depression and other disorders" (E. Bernstein D1). Whereas in the more immediate past, therapeutic intervention was reserved only for young children with overt behavioral problems, affluent parents at the turn of the twenty-first century pursue expert assistance for more mundane parent-infant issues, including "separation anxiety" (D1). Indeed, "There's a new push by doctors and therapists to identify children afflicted with anxiety disorders—even those as young as preschool age—and treat them early" (Petersen D1). Parents also pay for expert advice in combating obesity in toddlers (Wang "The War" D1), potty training, child-proofing the house, and specialized infant feeding and shopping services (Shin "A Coach" A1). Arlie Russell Hochschild describes a "commercialization of intimate life" at the heart of these pay-for-service technologies of child optimization (1–322).

The push to diagnose and treat very young children for anxiety and learning disorders reflects not only a commercialized culture of optimization, but also greater societal surveillance and pathologization of biopolitical differences. The biopolitical continuum between the normal and pathological child was fragmented by the child-guidance movement of the early part of the twentieth century, as discussed in chapter 2 (see also Nadesan *Constructing* 103–25). One of the first instances of the medical pathologization of children previously regarded as "normal" can be found in the early 1970s-era disorder of "minimal brain dysfunction" (Reinhold 27). Minimal brain dysfunction was a catch-all term for mild "learning disorders," but their unifying and characterizing feature was often described as attentional difficulties. As will be explained further in chapter 4, 1970s-era press stories of the disorder, described eventually as Attention Deficit Disorder (ADD) and Attention Deficit Hyperactivity Disorder (ADHD), included accounts of a seemingly miraculous new drug, Ritalin (generic methylphenidate), represented as capable of effectively and asymptomatically managing the symptoms of the disorder. Popularization of information about learning disorders such as ADD, later ADHD, and milder autism spectrum disorders (such as pervasive developmental disorder not otherwise specified [PDD-NOS] and Asperger syndrome) heighten parental anxiety about offsprings' developmental trajectories while responsibilizing parents for monitoring and remediating any perceived deficiencies or developmental delays.

Psychiatric drugs are often offered as a governing technology for managing risky children and children seen as at risk. Middle- and upper-class children seen as at risk for learning disabilities, anxiety, or depression, are referred to private physicians and psychologists who often, but not always, use medication to combat symptoms believed to interfere with learning, socialization, or appropriate conduct. For example, U.S. government data indicate that "prescriptions of antidepressants for all reasons rose to 42.6 instances per 1,000 office visits by children in the period of 1998 through 2001" (McGough D2). The 40 percent increase in bipolar diagnoses in children and adolescents from 1994 to 2003 led to a fivefold increase in prescription rates for powerful antipsychotics, particularly Risperdal (generic risperidone) and similar agents (Carey "Bipolar"; Harris "Use of Antipsychotics"). At least 25 percent of risperidone use occurred in children under eighteen years of age (Armstrong "Children's Use" B1). Use of drugs is increasingly replacing psychosocial therapies for all age groups because of health insurance reimbursement policies (Wang "Mental-Health" D5) and because of the drugs' relative ease of implementation.

Choosing the right medication for children is often challenging for medical providers. Psychiatric and psychological authorities often have a difficult time fitting children's symptoms into specific diagnostic categories. They also have a difficult time predicting how drugs developed for adults will impact children. Considerable experimentation often occurs when children are prescribed psychiatric drugs. Consider the following story reported in the *New York Times*:

> Paul was a gifted reader, curious, independent. But in fourth grade, after a screaming match with a school counselor, he walked out of the building and disappeared, riding the F train for most of the night through Brooklyn, alone, while his family searched frantically. It was the second time in two years that he had disappeared for the night, and his mother was determined to find some answers, some guidance. What followed was a string of office visits with psychologists, social workers and psychiatrists. Each had an idea about what was wrong, and a specific diagnosis: "Compulsive tendencies," one said. "Oppositional defiant disorder," another concluded. Others said "pervasive developmental disorder," or some combination. Each diagnosis was accompanied by a different regimen of drug treatments. (Carey "Parenting as Therapy")

In 2005 approximately 1.6 million children and teenagers were given at least two psychiatric drugs in combination even though the underlying causes of the symptoms targeted for pharmacological governance were not well understood (Harris "Proof" A1). Although pharmaceutical governance aims at normalizing and optimizing children's development, it entails a kind of biopolitical sovereignty through the process of labeling and therapeutic psychopharmacology (Nadesan *Governmentality* 170–71). It also offers opportunities for capital accumulation by pharmaceutical companies and medical authorities alike. Sales to children under eighteen years of age accounted for 15 percent of total antipsychotic sales in 2008 (Armstrong "Children's Use" B1, B10). Dr. Joseph Biederman, a child psychiatrist at Harvard University who has promoted use of the bipolar diagnosis for children, received $1.4 million in outside income from pharmaceutical companies who produce antipsychotics (Harris "Research Center"). Publicity about this unrecorded income in the media renewed concerns about the role of market factors in shaping perceptions and treatment of childhood deviance.

Many parents willingly adopt pharmaceuticals because they themselves have adopted the premises of biological selfhood and view emotional states as epiphenomena of chemical states (see Rose *Politics* 140). Judith Warner notes in her *New York Times* editorial that use of pharmaceuticals such as Ritalin has gone beyond remediation of perceived deficiencies, as "off label" uses have expanded to include enhancement of cognitive abilities by individuals seeking to outcompete their colleagues/peers (Warner "Living"). In essence, pharmaceuticals have been adapted to help leverage individual competitiveness in hypercompetitive educational and economic environments.

However, some parents are now seeking other means to treat their children's supposed deficiencies, even while accepting their purported biological bases. This passage from a *New York Times* article, "Parenting as Therapy," illustrates the shift in thinking:

> In recent decades, psychiatry has come to understand mental disorders as a matter of biology, of brain abnormalities rooted in genetic variation. This consensus helped discredit theories from the 1960s that blamed the parents— usually the mother—for problems like neurosis, schizophrenia and autism.
>
> By defining mental disorders as primarily problems of brain chemicals, the emphasis on biology also led to an increasing dependence on psychiatric drugs, especially those that entered the market in the 1980s and 1990s.
>
> But the science behind nondrug treatments is getting stronger. And now, some researchers and doctors are looking again at how inconsistent, overly permissive or uncertain child-rearing styles might worsen children's problems, and how certain therapies might help resolve those problems, in combination with drug therapy or without drugs. (Carey "Parenting as Therapy")

Cognitive therapy, psychotherapy, and intensive behavioral modifications have all emerged as approved expert strategies for treating children at risk. Implementation of these programs may not be covered by parents' health insurance (if insured), in contrast to psychiatric medication. Moreover, these types of programs are very time intensive and require extensive parental effort (e.g., as a parental technology of the self[3]) and critical examination of parenting techniques by self and expert authorities. In effect, parents must be able and willing to subject their parenting to intense surveillance and critical transformation.

Parents' willingness to think about their children's learning or behavior difficulties in terms of biological factors can be explained by many forces, but the circulation of brain imaging and genetic studies have surely played a role in conditioning their receptivity. Cognitive psychology's alliance with brain science may have contributed to parental receptivity by establishing linkages between familiar ideas, such as cognition, and new ideas, such as "synaptic growth and pruning." Therefore, while the cognitive brain science literature reinforced the idea that social environments shape intellectual capacity, it also directed public attention to the corporeality of the body. Brain science's space of visibility, after all, was the brain. Although this corporeal space was represented using the vocabulary and constructs of a cognitive psychology seeped in artificial intelligence, it nonetheless alerted the public to the importance of anatomy and physiology in shaping aptitudes and personal destiny. As biological accounts grew more common in the popular press in the 1990s, one sees a shift *away* from accounts that emphasize environmental inputs *toward* an approach that emphasizes biological predispositions. That shift will be addressed further in chapter 4, which explores how this focus on corporeality eventually directed attention away from environmental mediators and toward inborn mechanisms.

Having reached this point in the discussion of middle- and upper-class technologies of the self, the chapter will now turn more directly to those children who are widely perceived as "at risk" for failing to achieve their potential. At-risk children of the early twenty-first century, like those of the early twentieth century, are perceived as at risk for delinquency, dependency (poverty), degeneracy (e.g., drug abuse), and—most recently—for poor health. The following discussion will focus on those logics of government and disciplinary technologies that are promoted by child-welfare authorities, policymakers, and religious figures as most effective for "saving" at risk children. This discussion framework should not be construed as deliberately minimizing the efforts of poorer families to optimize their children by engaging in many of the same technologies of the self pursued by more affluent populations. Rather, the point is to demonstrate the disparate focus that emerges in public discourse and policy discussions about children of the lower socioeconomic classes.

GOVERNING RISKY AND AT-RISK KIDS IN THE EARLY TWENTY-FIRST CENTURY

Equality has rarely, if ever, served as an organizing objective of liberal logics of government. Although liberal social welfare logics favor risk socialization and mutual risk exchange at the level of the nation (M. Cooper 7), implementations of social welfare technologies of government have always been limited in the United States and have been constrained by idealized notions of market autoregulation. Federal social welfare programs implemented in the 1930s and 1940s in the United States were very limited in terms of the scope of their applications and sometimes met state resistance in implementation. Social welfare activism in the 1960s and 1970s generated a strong public backlash among those who resisted collectivization of societal risk.

Neoliberal, (neo)conservative, and libertarian critics believe inequality to be a natural condition of society. Indeed, as Frum explains, inequality is neither a problem of government nor a social dilemma from these perspectives:

> As long as there exists equality of opportunity—as long as everybody's income is rising—who cares if some people get rich faster than others? Societies that try too hard to enforce equality deny important freedoms and inhibit wealth-creating enterprise. Individuals who worry overmuch about inequality can succumb to life-distorting envy and resentment.

Frum points out that widening inequality of opportunity hurts conservative support but cautions that "equality in itself never can be or should be a conservative goal" because inequality is necessary for competition. Frum acknowledges that conservatives ought to combat *gross* inequality because it undermines competition. However, this acknowledgment does nothing to lessen conservative belief in the fundamental value of inequality in promoting the competition presumed as critical for efficient markets. Therefore, any efforts to combat gross inequalities must aim at fostering increased competitiveness, enabling individuals to compete in the market.

(Neo)conservatives and neoliberals argue that social welfare policies that shift risk to society away from individuals encourage a culture of poverty and reward lack of personal initiative. Social welfare policies are therefore perceived as market distorting because they allow individuals to avoid entering the market or they fail to deliver the competitive value orientation necessary for market success. For instance, many social conservatives harbor suspicion that government programs such as Head Start that are designed to enhance poor children's school readiness are inadequate for redressing the contaminating influence of the culture of poverty (or heritable inferiority). Moreover, conservative religious groups decry the secularism of government social welfare programs, arguing that they undermine moral accountability and personal initiative. Neoliberals simply reject social welfare policies as impinging on market freedoms by dampening "natural" competition and argue that government spending shifts resources away from markets to inefficient government bureaucracies.

Consequently, "neo" and traditional conservative logics of government converged in the last three decades of the twentieth century to transform public discourse about the government of poor and "at risk" children. This transformation in discourse has resulted in profound policy shifts. The discourse of equality and justice has been replaced by a discourse of personal responsibility, accountability, and punishment (see Giroux "Beyond" 587–620). As explained in chapter 2, the shift in formal governmental policy was achieved with President Clinton's welfare reform, President Bush's ownership society, and the institutionalization of faith-based initiatives (see Grossberg). This section next focuses on how these shifts in governmental discourse and logic have impacted perceptions and technologies of government of poor children's intellect and career opportunities.

Children, Poverty, and Biopolitical Divergences

Erosion of U.S. societal support for social welfare policies aimed at children can be contextualized by demographic data. Gallup polls reported in 2008 that only 29 percent of American adults have children in the age range of grades K-12 (http://www.gallup.com/poll/1612/Education.aspx). Consequently, by their sheer absence, children may lack significance as a societal concern for the majority of U.S. adults. Additionally, white America may reject social welfare logics given the growing diversity of America's children: in 2008 "racial and ethnic minorities" accounted for 43 percent of U.S. citizens under twenty years of age (Roberts). African American and Hispanic children are statistically more likely to live in poverty (Fletcher "1 in 4" A1), although the 40 percent of U.S. children living in or near poverty represents all ethnic groups. Still, only 10 percent of white, non-Hispanic children lived in outright poverty in 2008, compared with 35 percent of black children and 28 percent of Hispanic children (Federal 14). The median income for white households in 2005 was $50,622, for black households it was $30,939, for Hispanic households it was $36,278, and for Asian households it was $60,367 (Ohlemacher A3). Despite the circulation of these statistics, poor and minority children are typically invisible to affluent white populations outside of media depictions.

As explained in chapter 2, U.S. populations often live in areas segregated along economic and ethnic lines, creating very different childhood conditions, not simply in the home, but also in neighborhoods and schools. The divergence in childhood experiences along economic lines begins with prenatal care and nutrition. Poor populations are at greater health risk according to the statistics created by social welfare advocates. They have less access to health care, are more likely to smoke, and are more likely to be overweight and diabetic than more affluent populations (Isaacs and Schroeder 1137–42). Not surprisingly, poor populations have more health risks during pregnancy. College-educated mothers are likely to devour self-help literatures dedicated toward optimizing pregnancy outcomes and visit gynecological and obstetricians during their first weeks of pregnancy. Their work conditions often entail less physical and psychic stress than the demands of low-autonomy work experienced by working-class populations. Poor mothers are less likely to have health insurance and receive prenatal care (Wertz and Wertz 270). Food choices linked to healthy pregnancies by medical authorities are more expensive and may be less available to poorer populations as more lower-income Americans face food insecurity and lack access to well-stocked grocery stores (Williamson A1). Not surprisingly, the poor have less nutritionally balanced diets (Federal 18). These factors help explain why low-birth-weight rates have increased for singletons (Federal 60) and infant mortality rates have increased in the south (Eckholm "In Turnabout"). These adverse circumstances deprive less-affluent children of the biosocial environmental inputs believed to optimize developmental outcomes.

Early life experiences diverge significantly from their wealthier counterparts for the 40 percent of American children who live in or near poverty. Poor children are more likely to be subject to urban and industrial contaminants than affluent children and are also more likely to live in homes where someone smokes (Federal 30–31). For instance, 17 percent of black non-Hispanic children had blood level levels at or above 5 µg/dL in 2001–2004 (Federal Interagency 33). Exposure to urban pollution has adverse cognitive and cardiovascular effects. Recently, researchers linked children's measured IQ scores with prenatal exposure to urban air pollution, primarily from vehicle exhaust (Tanner "Kids" A1). The researchers claim their study documents the adverse, long-term effects of pollution. Poor inner-city children also risk high exposure to pesticides known to impact neurological development. One study conducted during the 1980s by the Environmental Protection Agency (EPA) found that 50 to 80 percent of Hispanic populations in San

Antonio and Houston, Texas, had residues of pesticide neurotoxins, especially Dursban, in their urine (Vallianatos). Finally, children who live in poor areas are more likely to suffer from health conditions, such as asthma, that adversely affect their daily activities (Federal Interagency 168, 171).

The 40 percent of children living in or near poverty are likely to have very different social experiences during childhood as well. For decades, researchers have stressed class-based differences in child-raising techniques. In the past, researchers linked observed differences either to a culture of poverty or to social deprivation. The distinction between these formulations was somewhat vague. Today, some researchers attempt to avoid pathologizing and stigmatizing the poor's parenting techniques. Instead of documenting the reproduction of poverty through poor parenting, contemporary researchers document how class-based parenting styles produce different orientations toward the world and different skill bases. Other researchers take a different approach by examining how the physiological and social stresses associated with economic marginalization impact children physiologically, limiting their opportunities. Both of these approaches attempt to provide environmental explanations for the persistence of poverty without pathologizing a culture of poverty/deprivation. These approaches can be regarded as relatively progressive in that they decenter a culture of poverty; although they fail to examine the economic conditions of possibility for poverty.

Annette Lareau's 2003 *Unequal Childhoods: Class, Race, and Family Life* argues that parenting styles diverge around class. She describes a "concerted cultivation" of children by middle-class parents (2), encompassing the technologies of the self described earlier in this chapter, which produce children as self-confident and entitled, but also armed with the communicative skills necessary for negotiating with adults. In contrast, working-class parenting promotes "the accomplishment of natural growth" (3) characterized by more authoritarian parenting, but also more day-to-day autonomy for children. Lareau suggests that these differences put middle-class children at an advantage in the context of middle-class social institutions. In contrast, working-class children, who are more likely to "gain an emerging sense of distance, distrust, and constraint in their institutional experiences," are at a comparative disadvantage (3). In other words, working-class children may lack access to the "social capital" (in addition to material capital) required for succeeding in middle-class institutions, including college and the corporate workplace.

Research has extended Lareau's findings to the divergent class-based child-care experiences as well (Nelson and Schutz). Welfare reform and the economic necessity for two-income families have together increased demand for child-care services among all sectors of the population. Nearly 70 percent of American women with preschool-aged children work full or part time (Naik). Affluent populations have more access to high-quality and expensive child care characterized by low student-caregiver ratios and credentialed teachers. These environments are most likely to draw upon child-care norms consistent with "concerted cultivation" and emphasize academic preparation (Nelson and Schutz). In contrast, poorer populations struggle to find affordable child care. The Rand Corporation report cited earlier in this chapter found significant class differences in the nature and quality of early childhood care and preschool education (Karoly).

The federal government provides some support for child care for low-income families through Head Start and through the Administration for Children and Families' (ACF's) Child Care and Development Fund (U.S. Department of Health and Human Services; http://www.acf.hhs. gov/programs/ccb/research/index.htm). However, the majority of families who are poor or near poverty struggle to pay for child care: in 2001, 40 percent of poor, single, working mothers paid 40–50 percent of their cash income for child care (H. Matthews). Inability to pay the high costs of licensed child care drives some parents to rely on home-based providers who may fail to meet minimal safety or staffing requirements expected of licensed caregivers. Children receiving poor-quality child care experience more neuroendocrine stress reactions than those receiving higher-quality child care (Gunnar 208–11).

Researchers who pursue environmental explanations for the reproduction of poverty have recently linked children's stress levels to outcomes such as cognitive development. Bruce McEwen attempts to quantify the stress of poverty by deriving it from levels of stress hormones (Stein "Research Links" A6). McEwen's model presumes that years spent in poverty correlate with higher stress levels and that stress levels are inversely correlated with a child's working memory. Thus, the model concludes that chronic poverty threatens children's cognitive functioning. Unlike the culture of poverty and social deprivation approaches, this framework explicitly views the conditions of poverty as producing physiological and psychological stress (e.g., from lack of financial resources, poor living conditions, etc.). This stress in turn produces marital discord and depression for parents, which exacerbate children's stress (Conger, Conger, Elder, Lorenz, and Simons 541–61). Poor children are also stressed by the isolation they may experience as their parents labor long hours (for poor wages) and by their cultural isolation from a society that worships affluence (Conger et al.). Researchers adopting this perspective argue that poverty cannot simply be eliminated by "better parenting."

Environmental accounts of the reproduction of poverty see affordable, quality child care as a critical social policy issue because of the importance accorded to the critical stages of infancy and early childhood (described earlier in this chapter). Better-quality child care, it is argued, can prepare children for school and can reduce children's and parents' stress levels. Yet neoliberal and conservative critics maintain that expanded public support for early child care and preschool represent market interference and unwarranted government expansion. For example, neoconservatives claim that Head Start attendance produces minimal gains for low-income children and that what gains can be measured diminish across time (Naik). The American Enterprise Institute published a report in 2007 quantifying the costs per child for early childhood education and care, implicating that Head Start was costly and market distorting (Besharov, Morrow, and Myers). The moneyed public generally lacks sympathy for poor, working parents' plight, seeing child care as a personal responsibility and liability.

Advocates for expanded government support for child care have responded by adopting neoliberal problem-solution frames when arguing that government support facilitates mothers' capacities to work and ultimately reduces government expenditures on the long-term costs of children. A U.S. government report published in 2000 argued for increased social welfare support for child care, given a market failure to provide high-quality care for the majority of young children, particularly infants and toddlers. Defining market failure as "a situation in which a market left on its own fails to allocate resources efficiently," the report observed that parents lack adequate information to make informed child care decisions because the market is made up of small providers, and considerations of costs, time, and access may adversely affect decision making, particularly for poor families (Vandell and Wolfe). Despite this market rationale for greater social welfare involvement in child care, little to no effort has been made since the beginning of the millennium to produce infrastructural supports for good-quality early childhood care (Gallagher and Clifford).

The experience of being poor as a child in a media-saturated popular culture that celebrates consumption and wealth is one of potential marginalization and shame. Poor children risk rejection by their peers, particularly when they live in close proximity to wealthier populations, because they lack the signifiers of conspicuous consumption defining youth culture. Fear of rejection or marginalization shapes childhood behavior in complex ways. Children may compensate overtly or retreat into shyness. Children eligible for additional supports may reject them, fearing stigmatization, as illustrated by many children's rejection of free lunches (see Pogash). Children's desires to fit in through material displays (electronic toys, expensive attire, cell phones, etc.) may very well exacerbate working parents' economic anxiety.

Poor children within the United States face declining social mobility as their parents struggle to make ends meet in an economy that is cutting jobs and benefits for the majority of workers. Otherwise-stable working-class children face homelessness when a parent loses a job or becomes ill. As David Shipler writes, "Poverty is like a bleeding wound. It weakens the defenses. It lowers resistance. It attracts predators" (18). High-interest loans, unscrupulous employers, and outright fraud exacerbate the poor's vulnerability to utter destitution. As will be discussed in the last chapter of this book, the economic meltdown that began in 2007 has plunged many more children into utter destitution. What is left of the U.S. safety net after neoliberal "reforms" offers little assistance for those who search in vain for nonexistent jobs (see DeParle "Slumping Economy"). Escape in the form of drug addiction and alcoholism can seduce a small sector of the population, exacerbating marginalization and societal contempt while creating living hells for children.

Statistically, in 2007 there existed more than a 40 percent chance that the children born of the bottom fifth of wage earners would remain mired in that same bottom quintile ("The Land of Opportunity?"). These children are increasingly cast as an economic burden on the nation. Their lack of technocratic education renders them "anachronistic" within neoliberal fantasies of economic innovation based on technology and speculative bio-economics. More recently, an alarm has been raised about the future health costs of today's children, since so many experience poor health and obesity. The economic security of the nation is seen as at risk as a result of future skyrocketing health costs.

Health Risks and the Burdens of Obesity

Skyrocketing health care costs capture national headlines weekly. The ranks of the uninsured grow as employers shift the responsibility of health insurance to employees. Low-income workers are particularly likely to lack health insurance. Federal and state expenditures for health coverage for poor populations have slowly expanded (e.g., SCHIP), providing a patchwork of health programs that help many facing illness. However, they typically fail to deliver effective preventive health care, and many Americans still lack any coverage at all (Joshi A2). Fearing unsustainable growth in health-care costs, public officials and health authorities have sought to govern the nation's wellness. Contemporary health government strategies aim to prevent adverse health outcomes by encouraging adoption of "healthy" individualized technologies of the self.

The poor and near poor have been targeted by these campaigns because of their apparent "unhealthy" lifestyles, as illustrated in this *New York Times* article:

> Many risk factors for chronic diseases are more common among the less educated than the better educated. Smoking has dropped sharply among the better educated, but not among the less. Physical inactivity is more than twice as common among high school dropouts as among college graduates. Lower-income women are more likely than other women to be overweight, though the pattern among men may be the opposite. (Scott)

Lower-income children's greater "risk" for obesity has emerged as perhaps the most critical public health menace in this discourse of prevention. Obesity has been identified as a public health menace in large part because of its linkages to diabetes, a disease estimated to have cost the nation $132 billion in 2002 alone (Kleinfield A1). Diabetes is a complex disease mediated by many factors, but high body weight is a known and measurable risk factor.

Contemporary efforts to govern poor children's weight illustrate how complex social dynamics are rendered governable in neoliberal formulations that individualize responsibility. "Eating right" and "exercise" are the primary weight management strategies promoted in public health campaigns (e.g., see "CEO Council on Health Care"). Schools and family pediatricians are encouraged to monitor children's weight and enroll overweight children in parent-implemented weight-reduction programs (Chaker A1; Tanner "Pediatric" A1). However, these seemingly reasonable prescriptions elide the complex reasons for the poor's relative unhealth. A brief look at some of these reasons challenges the efficacy of individualized solutions.

Poor children who lack access to costly and more difficult-to-obtain fresh vegetables are more likely to eat calorie-packed, but low-nutrition, food. Poor children who live in urban areas also have fewer opportunities for safe, pollution-free, outdoor play. Additionally, new evidence suggests that prenatal exposure to pollution increases a toddler's body mass index ("Programmed Obesity"). The poor are disproportionately subject to pollution. Finally, all parents, not simply poor parents, struggle against seductive food marketing of soda and sugary, fatty foods to children. In 2006 $492 million was spent in the United States on advertising for carbonated beverages (Marr D1). A total of $1.6 billion was spent on food marketing, mostly for fast food, cereal, and sodas in the same year (Marr D1). These complex synergies shaping weight and health defy simple, individualized solutions, yet the synergies' erasure in health policy discourse casts the poor as fully responsible for their health "deficiencies."

Michael Marmot's *Status Syndrome* contends that inequality and barriers to opportunity are alone capable of producing poor health outcomes. His approach locates health risk in societal structures and processes, rather than in individual behavior and choices. Poor populations, within this framework, suffer poorer health because of their occupational and living conditions. These conditions limit opportunities for healthy living while also encouraging a psychology of despondency (*not* dependency) that may further erode healthy choices. From this point of view, the unhealthiness of an increasing sector of the nation's children points directly to gross structural inequalities rather than to a lack of personal discipline or a culture of poverty that instills unhealthiness.

Global health indices are further proof of these structural inequities in the United States. The United States spends approximately $2.3 trillion on average a year for health care. Aggregate U.S. health spending is approximately twice that of other industrialized nations, figured per person ("CEO Council on Health Care"). Yet a 2007 UN survey ranking child welfare listed the United States last among developed nations surveyed in health and safety, based on vaccinations for childhood diseases, infant mortality, accidents, and death from injuries before age nineteen (McHugh A17). Inequalities in access to health care and healthy living conditions help explain why African American children have the worst asthma outcomes among population groups in the United States (Akinbami, Moorman, Garbe, and Sondik S131-S145). Structural, systemic conditions shaping American's children's poor health fail to capture media headlines or public policy responses because they cannot be figured into contemporary formulations for measuring and hedging risks.

School Segregation and the New Biological Determinism

White, middle-class children are most likely to live in outer suburbs with better-funded public schools and more opportunities for extracurricular activities and sports.[4] They increasingly live in gated communities. Affluent white children who live in major cities often attend private schools. Jonathan Kozol describes the racial segregation of schools in major U.S. cities:

> In Chicago, by the academic year 2002–2003, 87 percent of public-school enrollment was Black or Hispanic; less than 10 percent of children in the schools were white. In Washington, D.C., 94 percent of children were Black or Hispanic; less than 5 percent were white. In St. Louis, 82 percent of the student population was Black or Hispanic; in Philadelphia and Cleveland, 79 percent; in Los Angeles, 84 percent, in Detroit, 96 percent; in Baltimore, 89 percent. In New York City, nearly three quarters of the students were Black or Hispanic.

Segregation of American schools and residential areas blinds affluent Americans to the conditions and experiences of the growing segment of America's children who live or near poverty.

School segregation along lines of class and race has been exacerbated by neoliberal logics of government. As already established, efforts to divest the state of its paternalistic responsibility for citizens entail privatization of public pastoral (but secular) institutions. However, the impracticalities associated with privatizing public education have forestalled states' full-scale abnegation of their educational responsibilities. Instead, neoliberal states—such as the United States, Great Britain, and Australia—have mandated that public educational institutions adopt the prominent techniques associated with neoliberal regimes, including budget disciplines, accountancy, and audits (see Hill). In the United States, this push can be understood in relation to the neoliberal desire to increase public institutions' financial accountability by imposing upon their operations the discourse and practices of economic viability and market competitiveness (see Hursh). Moreover, neoliberal regimes of accountability have been promoted as a means for improving U.S. educational standing globally in light of data placing American students well behind Asian and European counterparts in the areas of math and science.

High-stakes testing has been adopted by many states as a strategy for increasing the visibility of educational processes and for disciplining schools that fail to achieve standards (see Graham and Neu 295–319; Hursh). In 2002 President Bush signed into law the No Child Left Behind Act of 2001, establishing standards for schools, including penalties for those that fail. The controversy surrounding the legislation stemmed mainly from the unfunded mandate of the legislation, rather than from its objectives to enforce transparency and accountability. When it was passed, only 13 percent of U.S. black eighth-grade students were judged proficient in reading; by 2005 the number had dropped to 12 percent (Tough). Although various biopolitical authorities, including philanthropic organizations and university researchers, explain test score gaps according to differences in school funding and the challenges of educating economically disadvantaged children, public policy simply attempts to link school-based test scores with fiscal rewards and punishments.

Pauline Lipman writes that school accountability policies and high-stakes testing produce a spectacle of discipline, as poor schools populated mainly with minority students are publicly shamed. Schools that fail to meet accountability standards are publicly chastised and are subjected to "a panoptic order of intensive monitoring and surveillance" (89). In the state of Arizona, such schools are publicly labeled as "failing." As Lipman explains, the public spectacle of school disciplining shifts blame to teachers and students while failing to address the conditions of possibility for educational outcomes. Underfunded schools, crowded classrooms, overworked and exhausted parents, and stressed children disappear in the politics of school accountability. For instance, the *Wall Street Journal* interviewed a panel of business authorities in an article titled "Failing our Children" in 2008. The contributors agreed that lack of national standards and accountability measures for teachers and students were the primary obstacles to improved educational quality. Pay-for-performance models were viewed as an important strategy for improving educational outcomes. The dilapidated state of so many U.S. schools was invisible in this policy dialogue because the framework utilized invoked problem-solution frames that centered on personal agency and responsibility.

At the K-12 level, local media and state school board efforts to publicize school "successes" and "failures," largely in relation to standardized test scores, have increased pressure on local school districts to demonstrate competence and competitiveness. This increased pressure, coupled with the partial privatization of schools through expansion of the charter school system in many states, creates opportunities for educational entrepreneurs to market all manner of educational services, from testing services and educational software to discipline and character education programs.

Educational and discipline programs selected by more affluent districts tend to emphasize development of students' capacities for self-governance and enterprise (see Nadesan "The Make Your Day") in contrast to the more disciplinary-type programs developed for, and adopted by, schools in lower socioeconomic areas. James Marshall suggests that a particular form of knowledge emerges from the convergence of enterprise values and disciplinary educational practices, which he describes as "busno-power" (3). Busno-power prioritizes technocratic forms of knowledge and learning, illustrated by fragmented and reductive learning processes emphasizing discrete units and skills requiring continuous, formal assessment (3). Busno-power, as an educational program and technology, finds greatest appeal in low-income schools pressed to demonstrate educational gains measured by standardized tests. Busno-power self-consciously incorporates "business" enterprise terms and value orientations (e.g., "quality" education measured by test scores and parental "customer satisfaction"), thereby appealing to policymakers drawing upon business models to evaluate education.

Increasingly, education is cast in technocratic terms, with both parents and educators agreeing that the curriculum should ideally prepare students directly for workplace needs, undermining philosophical commitments to the ideals of a classical liberal education (Marshall). While the type of knowledge required to prepare future white-collar workers is discussed and debated frequently in the press, little coverage is afforded to economic projections that future job growth will predominate in service occupations that do not require advanced education. More importantly, no public discussion exists on how to transform these positions into economically sustainable jobs for the individuals forced by necessity to labor within them. Instead, authoritarian school systems prepare students to accept these positions without questioning their conditions of possibility. Preparation occurs through school policies and practices stressing personal responsibility and discipline while emphasizing rote memorization and regurgitation of fragmented knowledge (i.e., busno-power).

In contrast to middle-class children with disabilities, the poor child who struggles with learning has no place at all in neoliberal logics of school accountability. Shipler illustrates the disparate experiences of poor children and middle-class children diagnosed with ADHD (248). Middle-class children receive specialized therapy and educational services in the home and at school. Poor children often receive neither and are ultimately constituted as public safety risks, as their diagnosis is correlated with future criminal tendencies and substance abuse. As will be explained further in the next chapter, the poor child's disability is increasingly cast as a heritable deficiency, while the middle-class child's disability is cast as a specific learning disability that can be governed through therapeutic protocols.

Most troubling, an entire "scientific" discourse has recently reemerged around the heritability of intelligence and school performance. This discourse particularly pathologizes poor, ethnic minorities. Beginning in the 1970s, claims about the social and educational implications of genetic heritability begin to circulate in the popular media, as illustrated by this 1977 article in the *Economist* affirming Hans Eysenck's assertion that earnings are determined 48 percent by genetic inheritance:

> Great society reformers created the impression, while not believing in it themselves, that all people could be tinkered into being equal. Outraged, Mr. Eysenck and others have campaigned to reverse that impression. They have been right to do so: offering equality of opportunity to the inherently unequal could, eventually, become a bitterly bad joke. ("Back to Genes" 11)

In the *Economist's* view, inequality is inherent, and societal outcomes merely reflect inborn differences.

Inequality became overtly racist (again) with Richard Herrnstein and Charles Murray's 1994 *The Bell Curve: Intelligence and Class Structure in American Life*, which argues that decades of research using intelligence tests prove that African Americans are unequal in intelligence to whites. The book was, of course highly debated, but the policy implications were clear for its adherents: social welfare environmental interventions for children are of limited value in ameliorating inherited intelligence. Charles Murray continued to advocate this position publicly, as illustrated by a series of editorials appearing in the *Wall Street Journal* in 2007 on the limits of education:

> Hardly anyone will admit it, but education's role in causing or solving any problem cannot be evaluated without considering the underlying intellectual ability of the people being educated. Today and over the next two days, I will put the case for three simple truths about the mediating role of intelligence that should bear on the way we think about education and the nation's future.
> Today's simple truth: Half of all children are below average in intelligence. We do not live in Lake Wobegon.
> Our ability to improve the academic accomplishment of students in the lower half of the distribution of intelligence is severely limited....This is not to say that American public schools cannot be improved. Many of them, especially in large cities, are dreadful. But even the best schools under the best conditions cannot repeal the limits on achievement set by limits on intelligence. (A20)

Murray's arguments may not have successfully swayed public policy, but they did erode public belief in the capacity of educational interventions to enhance equality of opportunity for poor children.

Academic interest in the relationship between genetics and educational achievement has recently grown. For instance, Miller, Mulvey, and Martin claim:

> The findings indicate that at least as much as 50 percent and perhaps as much as 65 percent of the variance in educational attainments can be attributed to genetic endowments. It is suggested that only around 25 percent of the variance in educational attainments may be due to environmental factors, though this contribution is shown to be around 40 percent when adjustments for measurement error and assortative mating are made. (211)

Researchers often represent the relationship between genes and educational achievement as mediated through the heritability of behaviors and conditions thought to impact learning, such as ADHD, conduct disorder, and learning disorders such as dyslexia (Caspi 281–301; Caspi et al. 851–854). This literature claims to avoid genetic determinism by incorporating social-environmental factors, which are explained as mediating the impact of heritability. For instance, parents' level of education is viewed as mediating genetic heritability in a study published in *Psychological Science* by Friend, DeFries, and Olson. These researchers conclude, "Genetic influence was higher and environmental influence was lower among children whose parents had a high level of education, compared with children whose parents had a lower level of education" (1). Still, the researchers admonish public educational policy for failing to acknowledge genetic limitations:

> Genetic constraints on learning rates are not recognized in the No Child Left Behind legislation, which requires that all children reach "grade level" (i.e., average) performance in reading and other academic skills by 2014, and assumes that this lofty goal can be met through appropriate education. A more beneficial policy would acknowledge genetic constraints on meeting a grade-level standard among some children with reading disability. (15)

The geneticization of educational attainment devolves risk in educational outcome upon the individual, thereby reducing the state's burden for compensatory education. This genetic logic challenges the entire framework of compensatory education by calling into question environmental adjustments' capacities to compensate for inborn limitations.

Research on the heritability of educational outcomes purports to acknowledge its limitations. Researchers acknowledge that heritability estimates of a trait operate at the level of the population and claim only to explain what percentage of variation in a trait found in a population is attributable to heredity. Yet a critical biopolitics holds that the very idea that genetic influences can be conclusively disentangled from environmental mediations is untenable when one acknowledges seriously the synergistic findings of contemporary genomics. What drives this research on the genetics of heritability given the irreducible synergy of biology and environment?

Clearly, research on the heritability of educational outcomes reflects the shift in risk away from social welfare strategies of management toward the responsibilization of individuals. Moreover, this shift in risk coincides with the financialization of societal risk-management strategies. If environment contributes only 40 percent of the variation in individuals' educational outcomes, policy makers can (purportedly) make informed, financially rationalized decisions about educational investments. In effect, this geneticization makes sense/cents within neoliberal problem-solution frames. Yet there is another dimension to this biopolitics that reinvents the legacy of classicist and racist eugenic attitudes toward poor populations discussed in chapter 2. Perhaps one of the greatest dangers of this new discourse is that it may harden public attitudes, enabling the disproportionate punishment and incarceration of poor children.

Discipline and Punishment

Although public awareness of educational inequalities exists, the fetishization of violence by the media has channeled educational coverage toward spectacles of school violence, drugs, and disorder (e.g., see Maeroff 3–9). Concerns about undisciplined and violent children in the schools are a common media theme; and yet, oddly enough, rarely, if ever, mentioned is the fact that nearly all cases of mass school shootings have been perpetrated by middle-class white males (Kimmel 500). Instead, childhood violence is represented most commonly as a ghetto phenomenon. Sharon Stephens points out how poor children are dehumanized in accounts that represent them as "unrestrained and undeveloped by the ameliorating institutions of childhood... children are represented as malicious predators, the embodiment of dangerous natural forces, unharnessed to social ends" (Stephens 13). Given widespread societal concerns about school violence and childhood criminality, policy makers and public opinion alike support school-based programs aimed at increasing school discipline, security, and the remoralization of children. These concerns and programs target poor children over their wealthier counterparts. This discussion examines some contemporary efforts to remoralize and discipline children, beginning with pastoral operations and moving to punishing ones.

Many schools today seek to remoralize students through older forms of pastoral governance. The Gallup Poll reports that one-third of Americans believe that the Bible is literally accurate (http://www.gallup.com/poll/1690/Religion.aspx). Therefore, it is not surprising that even Democratic politicians occasionally endorse bills allowing schools to offer Bible study, particularly as promoted in a new textbook produced by the ecumenical Bible Literacy Project, *The Bible and Its Influence* (Kirkpatrick). Growing overt religious influence in schools has resulted in some resistance (Banerjee), but the tendency for schools to "outsource" operations such as sexual education has created opportunities for religious proselytizing and has reframed formerly sterile pedagogical discussion of reproductive processes into moralizing discussions of abstinence and the dangers of sexual contamination (see Fisher B1; Weiss A17). Historically, black American teenagers have been the primary targets of efforts to control teen sexuality (Washington 189–215), although Latinos are increasingly targeted as well.

Conservative Christian pedagogical practices can converge with the precepts of neoliberal governance, including Patrick Fitzsimons's idea of the "autonomous chooser." As Fitzsimons explains, the model of the autonomous chooser implies a kind of Kantian "faculty of choice" in a context requiring perpetual response. The child's integrity, or wholeness, is conditioned and expressed by his or her faculty of choice. Conservative Christian pastoralism characterizes the child as choosing between subservience to God, on the one hand, and the pursuit of temptation, on the other. Poor children, tempted by unattainable marketplace goods, are expected to "choose" to respect property rights and official authority, even when these sovereign principles and persons seem to perpetuate the environment of dispossession experienced by marginalized children. Within the conservative imagination, failure to respect sovereign principles and authorities demands forceful discipline and punishment. Suspicion that poor and minority children will fail to "choose" wisely encourages surveillance networks within communities and schools aimed at preempting resistance.

Programs targeting "at risk" youths or youths from lower socioeconomic categories individualize societal risks while disciplining those youths targeted for surveillance. As Peter Kelly writes, the "discourses of youth at-risk seek to individualise the risks to the self that are generated in the institutionally structured risk environments of the 'risk society,'" thereby constructing such youths in terms of "individual pathologies and deficits" (23, 24). Although "at risk" discourses may reflect humanistic impulses to assist targeted children, these impulses are increasingly interpreted within an epidemiological framework of public risk. As Nikolas Rose explains, the biomedical model of risk and public health has been extended to encompass individuals who are viewed as "anticitizens" because of their purported incapacities for self-government (*Politics* 242). Such individuals are targeted for intervention in order to protect the public well-being, even if said individuals are victims of circumstances, including poverty, deprivation, mental illness, and so on.

Accordingly, in contrast to the programs developed for, and selected by, wealthier districts, educational practices adopted by less affluent districts seem aimed at increasing student visibility within, and compliance to, pedagogical disciplinary apparatuses. In this context, particular regions of the United States, especially those with high numbers of Christian conservatives, such as the southern Midwest and rural South, widely practice spanking in schools in order to punish children whose "choose" resistance. A 2006 *New York Times* article reports:

> The most recent federal statistics show that during the 2002–3 school year, more than 300,000 American schoolchildren were disciplined with corporal punishment, usually one or more blows with a thick wooden paddle. Sometimes holes were cut in the paddle to make the beating more painful. Of those students, 70 percent were in five Southern states: Texas, Mississippi, Tennessee, Alabama and Arkansas. (Lyman)

These corporal modes of discipline continue in twenty-one states despite statements against the practice by various "secular" biopolitical authorities representing the American Academy of Pediatrics, the National Association of School Psychologists, and the American Medical and Bar Associations (Lyman). Conservative advocates argue against secular authorities by invoking biblical authority, stating that the Bible condones corporal discipline and that the practice reduces school discipline problems.

The statistics on the use of corporal punishment in schools suggest that black and disabled students are subject to disproportionate disciplining (ACLU "A Violent Education"). A study of corporal punishment in Arkansas using the state's own educational statistics found that "the proportion of the 36,439 paddlings administered that involved an African American was over one and one-third times the proportion of student enrollment that was African American (31.19% divided by 22.75% reveals the problematic disproportion)" (http://www.neverhitachild.org/Arkansas/). In Florida's Duval County, nearly 80 percent of the paddlings went to black students who comprised 43 percent of the county's school student population (Garza). These data suggest that school administrators regard black students as more defiant or as less capable of personal self-governance than white students.

School authorities' willingness to use physical means to discipline lower-income children reflects complex dynamics, but societal stereotypes may help legitimize their use of force. Derogative social stereotypes about the "culture of poverty" are not simply acquired from the media but are actually promoted by some well-established educational consultants. One Texas educational consultant, Ruby Payne, travels nationally promoting her "Framework for Understanding Poverty," which "teaches the hidden rules" and mind-sets of economic class (11). Payne's book instructs middle-class teachers that poor parents are more likely to beat their children and verbally chastise them: "The typical pattern in poverty for discipline is to verbally chastise the child, or physically beat the child, then forgive and feed him/her" (23). Payne claims that individuals in poverty have a strong belief in fate and little belief in the capacity for personal change (23). In contrast, Payne offers readers the following measures of middle-class knowledge and capabilities:

> I know how to get my children into Little League, piano lessons, soccer, etc.
> I know how to properly set a table.
> I know which stores are most likely to carry the clothing brands my family wears.
> My children know the best name brands in clothing. (39)

Payne's stereotyped representations have been examined and challenged by academics, yet her ideology and pedagogical strategies continue to circulate widely (Shapira A1). The idea that the poor lack ambition, cultivation, and (consumer defined) taste, coupled with their purported conditioning to authoritarian parenting, contributes to repressive and condescending pedagogical attitudes.

The authoritarian, disciplinary framework that prevails in many U.S. schools echoes the Puritan belief that children must be beaten into submission. While the helping-hand therapeutic ethos is employed in affluent school districts, the discipline-and-punish ethos prevails within too many districts populated by lower-income students. Moreover, schools' increased use of zero-tolerance school discipline policies seems to be disproportionately targeted at black and Hispanic students (Lipman 4–5). A recent case illustrates the use of repressive police apparatuses to discipline minority children:

> When 6-year-old Desre'e Watson threw a tantrum in her kindergarten class a couple of weeks ago she could not have known that the full force of the law would be brought down on her and that she would be carted off by the police as a felon...The child was fingerprinted and a mug shot was taken. "'Those are the normal procedures for anyone who is arrested," the chief said. (Herbert "Six-Year-Olds" A17)

The indiscriminate use of police officers as school disciplinarians has attracted national attention, yet it continues (e.g., Herbert "Harassed" A1).

Lower-income disabled children and children who exhibit behavioral problems are particularly susceptible to disciplinary technologies. In many cases, public schools lack adequate resources to provide needed services to children with disabilities. Overcrowded schools subjected to inflexible regimes of accountability and faced with behaviorally difficult special needs children sometimes resort to physical restraint and seclusion. The news media increasingly report incidences of kids diagnosed with ADD and autism being traumatized by their mistreatment in public schools. Most recently, a thirteen-year-old boy hanged himself when forced repeatedly into a cell-like school seclusion room (Fantz). Disabled, stressed, and anxious children pose real problems for poorly prepared and understaffed teachers tasked with their education. Yet spectacles of children's death or suffering occasionally revealed in the popular media fail to change zero-tolerance policies and school underfunding. In addition, children with mild disabilities are also subjected to disproportionate bullying (e.g., see Barry). Disabled children without specialized educational protections in highly authoritarian environments may be more likely to be rendered vulnerable to bullying for a variety of reasons. Lacking the material resources, symbolic capital, or confidence to fight school administrators, their parents may be unable to provide them the same level of protections afforded the disabled children of affluent parents.

Kids Punishing Kids

Bullying among students in schools has become a "police" issue, as explained by a U.S. Department of Justice report titled "Bullying in Schools" (Sampson). Bullying is a police issue because it is linked to security. This report observes that "perhaps more than any other school safety problem, bullying affects students' sense of security" (1). The report emphasizes that bullying is pervasive, underreported, and damaging psychologically. Bullying is defined as

> *repeated harmful* acts and an *imbalance of* power. It involves repeated physical, verbal or psychological attacks or intimidation against a victim who cannot properly defend him- or herself because of size or strength, or because the victim is outnumbered or less psychologically resilient. (2)

The report distinguishes conflict between two individuals of equal psychological or physical strength from bullying. The idea of an imbalance of power is critical to the report's definition of bullying.

Bullying became a public policy issue after well-publicized instances of school shootings, including the Columbine murders. Reports of widespread school violence led to a number of research studies attempting to explain the sources of bullying behavior. In 2001 the National Institutes of Health (NIH) reported that bullying is pervasive in U.S. schools and that both victims and perpetrators are more likely than their peers to experience social isolation ("Bullying Widespread"). The U.S. Department of Justice report similarly attempts to identify individual traits or experiences that predict bullying behavior and concludes that children with controlling, punitive parents are more likely to bully. In additional, bullying behavior appears statistically more frequently in lower socioeconomic areas.

Both the Justice Department and the NIH reports *individualize* bullying in terms of personal characteristics. Individualization of bullying simplifies efforts to govern bullying behavior because it provides a framework for identifying kids likely to bully and kids at risk for being bullied. This framework therefore suggests that potential bullies can be subjected to targeted governance, including heightened surveillance and therapeutic intervention. However, this approach to bullying is more likely to identify instances of physical bullying than it is to identify psychological bullying. Moreover, nine years of research and intervention efforts on school bullying seem to have had little to no impact on the phenomenon.

The individualized, psychological framework used to understand and govern bullying no doubt limits the success of intervention efforts. One variable that recurs in variable-analytic approaches to predicting bullying is socioeconomic status. Children of lower socioeconomic status living in countries with greater socioeconomic gaps are more likely to be bullied (Merlo et al. 907–14). The stressors of economic marginalization no doubt play a role in the family dynamics that produce children prone to bullying. The rage produced by socioeconomic marginalization is directed toward targets whose perceived physical or social "deficits" render them vulnerable. Targeted governance of potential bullies by school or police officials, within this alternative interpretive framework, simply individualizes and psychologizes systemic faults produced by highly unequal, status-conscious societies. Furthermore, the enhanced surveillance and disciplinary environment experienced by socioeconomically disadvantaged kids in public schools may contribute to bullying and violence by kids against kids

Incarcerating Children

Youth violence and gang membership tend to headline media discussion of youth poverty in the United States despite recent declines in rates of serious crimes by youth perpetrators (Federal Interagency 46). Still, school violence and youth gang violence remain significant problems in some urban areas. Having already explored school violence, this section addresses societal strategies targeting youth gang membership and youth criminal activities.

In *Dead Cities: And Other Tales,* Mike Davis depicts the socioeconomic deprivation in Los Angeles County that makes gang membership enticing for young people who see little hope of overcoming their marginalization or achieving a decent standard of living through legitimate means. Drawing upon classical ethnographic accounts documented by Pruitt-Igoe, Steven Mintz suggests that inner-city youths' awareness of hostility directed toward them by middle-class culture may contribute to gang membership because it often provides employment opportunities, respect, and identity for children who have little to no access to these in mainstream society (352). Tragically, however, gang membership increases young people's susceptibility to homicide, as demonstrated by the escalating homicide rate by and of African American youths (Eckholm "Murders"). Gang activities may also reinforce and encourage violent social relationships.

Efforts to combat gang membership and violence over the last decade often involved arresting and detaining suspected gang members and prosecuting juveniles as adults, leading to stiffer sentencing. Since the 1990s, thirty-six states passed hard-line laws aimed at gangs (Moore "Gangs Grow"). These laws prohibit public gatherings of two or more individuals suspected of gang membership, create and maintain databases of gang members, use broad sweeps of suspects, and extend prison sentences for gang-related crimes. Economically disadvantaged youths are thus governed by a combination of enhanced police surveillance, legal criminalization of gang membership, and zero-tolerance juridical policies. Stiff sentencing for minor offences and three-strikes laws that incarcerate repeat offenders for life irrespective of the severity of crimes committed have greatly increased overall U.S. incarceration rates (Aizenman "New High" A1). Paradoxically, these strategies aimed at preventing and containing youth violence often involve institutional violence against youths.

Once children enter the juvenile court system, particularly minority and poor children, they are in danger of being sentenced to juvenile "boot camps," privatized facilities where they are subjected to remediation, often through strict disciplines and corporal punishment, as illustrated by the recent beating to death of a Florida boy at the hands of his boot-camp "instructors" (Goddard). In 2007 the following *New York Times* article described conditions within a Texas juvenile detention center:

> AUSTIN, Tex.—Juvenile detainees as young as 13 years old slept on filthy mats in dormitories with broken, overflowing toilets and feces smeared on the walls. Denied outside recreation for weeks at a time, they ate bug-infested food, did school work that consisted of little more than crossword puzzles and defecated in bags. After months of glowing state reports, the squalid conditions were disclosed on Oct. 1 by state inspectors at the Coke County Juvenile Justice Center in Bronte. They are another sign of the deep disarray of the Texas Youth Commission, the nation's second-largest, after Florida's, and most troubled juvenile corrections agency. (Moore "Troubles Mount")

As illustrated by this article, boot camps reinvent the abusive horrors of earlier nineteenth- and early twentieth-century detention institutions for young people.

Society contains its contradictions through incarceration. As Orlando Patterson reports,

> America has more than two million citizens behind bars, the highest absolute and per capita rate of incarceration in the world. Black Americans, a mere 13 percent of the population, constitute half of this country's prisoners. A

tenth of all Black men between ages 20 and 35 are in jail or prison; Blacks are incarcerated at over eight times the white rate.

The public white imagination, conditioned by at least two centuries of racist propaganda, conceives this tragedy as deriving from either the inborn deficiencies or the culture of poverty (supposedly) afflicting black America.

The role of political economy (e.g., high unemployment and poverty) in driving criminality, criminalization, and sentencing is swept aside by the compelling economic incentives of the prison-industrial complex, through which an entire nexus of private and public apparatuses have converged to exploit the bodies and labor of incarcerated prisoners. Dissent exists, as illustrated by a 2007 report by the NAACP Legal Defense and Educational Fund, which describes a school-to-prison pipeline that begins with underfunded and neglected schools but also includes "overzealous" discipline policies that remove students from school and rely on police officers to dispense punishment through the criminal justice system, blurring discipline and punishment (J. Johnson). "No tolerance," "zero tolerance," "public safety" campaigns have resulted in more youths being sent to detention facilities, profiting privatized juvenile detention facilities. Dissent against widespread incarceration of minors has been marginalized.

Privately owned and operated institutions have vested economic interests in juvenile incarceration, as do privately run prisons for adult populations. Poor rural areas rely on these institutions to provide work for otherwise unemployed citizens, dampening public outrage against the institutionalization of children in for-profit punishment centers. In 2009 the U.S. media began publicizing reports that approximately 2,000 juveniles had been sentenced to detention centers by judges who received "kickbacks" exceeding $2.6 million (Pilkington). The pervasiveness of fraudulent decision making in the juvenile justice system has not yet been investigated. Still, media publicity of the sickeningly abusive environments of juvenile detention centers and boot camps may eventually encourage states to adopt more pastoral approaches toward their risky children (e.g., see Moore "Missouri System").

Ironically, the cultural preoccupation with disadvantaged youths, gang violence, and street crimes obscures the crimes that most adversely impact the economic vitality of the nation. Just as the biopolitical statistics on bullying elide the less overt (i.e., physical) forms of bullying found among affluent students, so also do the biopolitical representations of the precursors of adult crimes among disadvantaged youths obscure the impact of white-collar crime. In 1949 Edwin Sutherland defined white-collar crime as "a crime committed by a person of respectability and high social status in the course of his occupation" (9). White-collar crimes range from outright fraud and embezzlement to health and environmental crimes perpetrated by businesses and corporate entities. The savings and loan disasters of the 1980s, the Enron and waste management accounting scandals in the 1990s, and the recent Bernie Madoff Ponzi scheme illustrate white-collar crimes, as does Peanut Corporation of America's knowing distribution of peanut products contaminated with salmonella. An entire genre of research documents that white-collar crimes destroy more lives and produce more economic costs to the nation than street crime, yet media outrage occurs only in the immediate aftermath of publicity about some recent instance. Yet in contrast with street crime, no widespread efforts are made to reduce white-collar crime by governing the social morality and business ethics of American's affluent children in K-12 education.

Military Service as "Salvation"

Many young men and women who emerge as risky adults in blighted urban areas have relatively few avenues of escape from the poor economic circumstances of their communities. Community colleges and public universities alike have become less affordable as a result of the de facto "privatization" of higher education (Dillon). One escape option that is often celebrated in the press is military service. The U.S. military promises career training to disadvantaged youths who lack the economic and symbolic resources to pursue college education. *Slate* magazine claims that the total military/defense budget for 2008 was $713.1 billion (Kaplan). U.S. empire building abroad to gain access to and protect resources requires a ready source of willing soldiers who are deployed globally. As chronicled by Michael Klare in *Blood and Oil*, overseas deployments are increasingly occurring in unstable regions troubled by guerrilla conflicts (1–247). U.S. soldiers police resources while Special Forces engage in clandestine activities aimed at keeping friendly officials in power. U.S. economic mercantilism offers employment for individuals who often have few other options available.

Military recruiting officers often have access to high schools and target students in lower socioeconomic areas. At staged school rallies, military service is represented as a good career choice for young people who have few opportunities. A new approach to recruitment is found in the growing numbers of high school military academies run by the U.S. armed forces that are popping up in low-income urban areas with deteriorating public schools (Roa). In addition to academics, these schools teach military disciplines. They therefore have considerable appeal for authorities who seek to control inner city life.

The American Civil Liberties Union (ACLU) issued a report in 2008 titled "Soldiers of Misfortune," claiming that the U.S. military recruiting practices target children as young as eleven years, mostly among poor and minority populations. Recently, however, the military developed a particularly persuasive recruiting tool designed to captivate nearly *all* American boys. In May 2002 the U.S. Army helped launch a video game titled "America's Army." The downloadable game was offered free of charge. It features a Mid-East urbanscape within which players can simulate warfare using virtual weapons replicating those actually used by the U.S. Army. Ubisoft, the multimillion-dollar publisher of the console version of the game, asserts that "America's Army" is the "deepest and most realistic military game ever to hit consoles," hoping that it gave players a "realistic, action-packed, military experience" (cited in Reagan). Violence is fetishized by computer games such as America's Army and Company of Heroes, supplementing the never-ending parade of "action-adventure" films and television series (e.g., Fox's *24*) that cultivate combative masculinity and promote Manichaean conflicts (see Rutherford 622–42). The U.S. military offers young men a socially acceptable vehicle for channeling aggression with the bonus of career development promises.

The young men and women who join the military and are unlucky enough to experience combat abroad discover their vulnerability to the real-world risks of war, as evidenced by brain injuries and posttraumatic stress (Dreazen A4). In 2009 the Veterans Administration estimated that 320,000 returning soldiers from Iraq and Afghanistan suffer from traumatic brain injury ("VA to Increase"). Veterans also discover their risk for unemployment. In 2005 military.com reported that the unemployment rate for veterans of the Iraq and Afghanistan wars was three times the national average ("Unemployment Rate"). Unemployment rates for veterans continued to rise through 2008 ("Increase"). The combination of posttraumatic stress and lack of economic opportunity may help explain the rising suicide rate among male veterans aged eighteen to twenty-nine (Zoroya) and the spike in domestic violence cases involving returned veterans (Alvarez and Frosch). For many returning veterans, the seductive recruiting promises and exciting adventures abroad have been revealed to be false promises and false fantasies of glorious war. With psyches scarred by violence and bodies ravaged by combat, America's former soldiers emerge as security risks in a new discourse promulgated by the Department of Homeland Security (DHS) that represents disaffected former soldiers as at risk for joining domestic terrorist groups (U.S. Department of Homeland Security "Rightwing Extremism").

The economic recession that began in 2007 may indeed encourage growth of radicalized groups of young men and women. The DHS report observes that

> the economic downturn and the election of the first African American president present unique drivers for rightwing radicalization and recruitment . . . Nevertheless, the consequences of a prolonged economic downturn—including real estate foreclosures, unemployment, and an inability to obtain credit—could create a fertile recruiting environment for rightwing extremists and even result in confrontations between such groups and government authorities similar to those in the past. (U.S. Department of Homeland Security "Rightwing Extremism" 3)

The report explains that right-wing militias groups in the United States are mobilized by the apparition of hordes of illegal immigrants and by a conspiracy aimed at establishing a "one world government" (5–6). These narrative themes echo those articulated by decades of science fiction fantasies, especially *Star Trek* and *Star Wars*, which define and promote combative masculinities.

CONCLUSIONS

The discourses, practices, and institutions of American society are shaped by contradictions that impact the governance of American children. The ideology of self-creation and economic opportunity seduces the upper and aspiring classes, leading to intense parental surveillance and pastoral cultivation of America's upwardly mobile children. Middle- and upper-class children are believed to be responsive to environmental inputs, and parents therefore practice a wide array of technologies of the self to ensure their children's future success. Parental and educational technologies used to cultivate affluent children are believed to foster an innovative and flexible U.S. workforce, capable of competing in a globalizing economy. However, new concerns about the role of the helping hand in producing an ethos of entitlement and egoism are gaining traction, presenting novel risks for the government of America's most affluent children.

Less economically advantaged parents also hope to optimize their children's opportunities, but these parents' lack of access to economic and symbolic capital can limit their capacities to provide the intense career preparation pursued by more affluent classes. Throughout much of the twentieth century, social welfare advocates tried to "assist" poor populations through parenting and compensatory education programs. Early childhood emerged in the post–World War II period as a critical temporal space for intervention. However, it was the affluent social classes who were most successful in implementing research findings as the project of remediation was adapted for middle-class children's optimization in the closing decades of the twentieth century. Simultaneously, the stressors of poverty stemming from low wages, poor living conditions, and underfunded schools were eclipsed in evolving neoliberal and neoconservative discourses that stressed individual responsibility and transformation. Today, social welfare logics have been eroded by the tide of neoliberal values, promises, and programs. America's affluent populations are more likely to view the nation's poorer children—upward of over 40 percent of the childhood population—as risky, rather than "at risk." Contemporary policing of risky children sheds child-saving logics in favor of logics deriving from public risk management. The differential treatment of poor children as risky encourages incarceration, particularly of African American children. Henry Giroux offers a "biopolitics of disposability" to capture these trends and effects shaping the lives of less affluent children ("Beyond the Biopolitics" 587–620; see also L. Grossberg 1–384). Evolution of this biopolitics and its transformation across twentieth-century governmental logics organize discussion in the next chapter.

CHAPTER 4

Biopolitical Sorting: Comparing Neoliberal and Social Welfare Problem-Solution Frames

Contemporary children are regarded as at risk for ill-health effects from a wide variety of environmental and biological perils. Environmental perils range from toxic substances such as lead and mercury to the nutritional deficiencies of high-fat, fast-food diets. Biological perils range from external threats such as contagious diseases to "inborn" ones such as faulty neurochemistry and defective genes. New technologies have been developed to reveal children's health risks even before they are born. Inborn perils such as cystic fibrosis can be detected in embryos through amniocentesis. Environmental risks to future offspring can be detected in mothers even before pregnancy. For instance, mothers' blood can be tested for heavy metals known to adversely impact development, such as lead.

In the culture of fear that accompanies efforts to securitize offspring, even necessary or ubiquitous commodities and lifestyle practices can produce risks to children's health. Too much television can purportedly stymie cognitive development or even cause autism, according to some concerned child activists (Wallis). Too much food, particularly the "wrong" kinds of food, produces obesity. Too much television and too much of the wrong kind of food might even cause type II diabetes. The risks to children's well-being proliferate as new links are established between children's health and their complex biosocial environments. Moreover, the parameters of normality in physical and mental health become ever more exacting as new technologies are developed for sorting childhood populations according to their exposure and susceptibility to biological and environmental risks. It is important, therefore, to examine how changing understandings and measurements of environmental and biological risk contribute to and reflect the changing problematics of the societal government of childhood.

Discussion will focus on two types of risk that are specific to the late modern period, environmental risk and genetic risk, by focusing on lead poisoning and ADHD. As an "environmental" disease, lead poisoning was both new and radical in the mid-twentieth century because it pointed to the role of the built environment in directly harming children's physiological and intellectual development. However, as shall be explained, efforts to remove this environmental threat to children's health were undermined by industry interests and by racist public attitudes. The history of public policy approaches toward lead poisoning therefore illustrates the politically contingent nature of biopolitical efforts to secure children's health. The second type of risk to children's health explored in this chapter, genetic risk, has recently captivated the attention of the national media, the National Institutes of Health (NIH), and privately sponsored biomedical research. The public has been slower to adopt the framework of genetic risk when considering children's health but is gradually adopting this paradigm through prenatal genetic testing and postnatal genetic screening for diseases. What is particularly interesting about genetic risk is that this type of health risk is modulating and occasionally eclipsing environmental explanations for childhood disorders such as ADHD. Accordingly, behavioral conditions that might previously have been regarded as stemming from environmental influences are slowly being reconfigured as gene-based or gene-linked disorders. Discussion begins by briefly sketching historical accounts of risks to children's health.

Children's Health Risks: From Morbific Environmental Influences and Degeneracy to Environmental and Genetic Risks

As explained in chapter 2, morbific environmental influences and inborn degeneracy were understood as the primary threats to infants' and children's health prior to the late nineteenth century. Communicable diseases such as typhoid and measles were known to be contagious, but Western science did not identity the role of bacteria and viruses in spreading diseases until the 1870s. Indeed, eighteenth- and early nineteenth-century inoculation against smallpox using cowpox material occurred without a clear understanding of either disease's mode of transmission. Therefore, historical efforts to prevent disease outbreaks from spreading primarily focused on establishing a *cordon sanitaire* that separated contaminated areas from uncontaminated areas (see Armstrong A New History 7–8). The sanitarians adopted this principle in household maintenance, producing a regime of domestic hygiene that included sanitation through cleanliness and ventilation. Their model of prevention through hygiene was eventually adapted to the "problem" of degeneracy and criminality as well. A brief explanation of these developments contextualizes twentieth-century approaches to understanding and managing children's health risks.

In 1883 the U.S. surgeon general of the Marine Hospital assumed responsibility for overseeing the nation's health (Kober 796). The responsibilities of the position involved establishing quarantines, monitoring sewage contamination of water, monitoring adulterations and milk contamination, controlling opium consumption, expanding vaccinations, and patenting proprietary medicine (795–99). As scientific understanding of disease vectors grew in the early twentieth century, the federal and state departments of public health also promoted personal hygiene practices such as hand washing to prevent interpersonal transmission of disease. Public schools helped disseminate newly developed hygienic practices by the early 1900s (see Sedgwick and Hough 132). By the early twentieth century, public health authorities were primarily concerned with combating the transmission of communicable diseases.

Chapter 2 explains how the regime of hygiene used to combat communicable diseases was adapted for the prevention of mental illness in the early decades of the twentieth century. Children were targets of a novel strategy for preventing criminality and madness. Children had been as viewed as mostly immune to mental illness until the second half of the nineteenth century (Walk 754–67). Once children were discovered to suffer from mental illnesses other than idiocy and fever-induced brain damage, psychiatric experts were prone to explaining their condition in relation to inborn degeneracy and constitutional flaws. Late nineteenth-century child savers suggested that the contagion of adverse environments posed by unsanitary environments and unsavory parents might also play a role, ultimately prompting the "preventive" ethos described in chapter 2. The resulting medical-hygienic complex sought to prevent infant mortality and child mental illness and delinquency through philanthropic and public services aimed at educating mothers on proper hygiene, feeding, and care of their children.

Early twentieth-century efforts to cultivate children's physical and mental health focused on communicable diseases, nutrition, and mothering. Contemporary concerns about "environmental health risks" simply did not exist until some years after World War II. For instance, in 1953 environmental health was defined exclusively in terms of unsanitary water and ineffective control of flies and insects in *Public Health Reports* (S. Smith 203). Early environmental health studies were beginning to link occupational exposures to lung cancer in the late 1950s (Haenszel 154), but research had generally not yet begun to explore the public's susceptibility to environmental hazards outside the workplace. The first article to appear in a search on environmental risk to public health in the research database JSTOR is titled "Reality and Perception of Environmental Hazards," published in 1964 (Van Arsdol, Sabagh, and Alexander 144–53).

Scott Frickel argues that political expansion of institutions concerned with broad-scale environmental risks to public health began in earnest in 1969 when the federal government passed the National Environmental Policy Act (NEPA), which initiated policy actions addressing biological and ecological impacts of synthetic environmental chemicals (185). NEPA mandated creation of the Environmental Protection Agency (EPA) and the Council for Environmental Quality. Also founded in 1969 was the National Institute of Environmental Health Science (NIEHS), which directs basic research on the effects of environmental factors on human health (185). Frickel explains how terms such as *genetic toxicology* and *environmental mutagenesis* came into circulation after 1966 (190).

Lead poisoning played an important role in mobilizing public concern about environmental hazards in the 1960s and 1970s.[1] Examining public discourse about lead poisoning is instructive because it demonstrates how an environmental hazard became constituted as a "risk" only when it was linked to broader sociopolitical concerns affecting middle-class populations. As will be discussed, lead poisoning was regarded as a rare occurrence in the first half of the twentieth century because medical accounts were limited to a few clinical case studies. Epidemiological methods had not yet been applied to studying environmental disease agents. It took decades of case studies of mostly poor lead-poisoned children before epidemiological methods were applied to study the wider prevalence of the problem. Several decades of epidemiological research eventually revealed *all* children as potentially at risk from lead in paint and fuel. Only then were laissez-faire attitudes against government regulation of lead overcome with regulatory controls aimed at protecting children.

Epidemiology studies whole populations in order to evaluate the frequency and distribution of disease and to measure risks for disease outbreaks. As explained previously, epidemiology is applied also to "problems" of mental health and crime. Public health authorities interested in the spread of diseases such as cholera, typhoid, and smallpox adopted quantitative methods of data collection and analysis in the mid-nineteenth century in order to promote "scientific" respect for their findings (Petersen and Lupton 28). Epidemiological data were mined for variables believed to reduce or increase risk of outbreak and transmission. For example, living in basements was correlated with propensity for particular contagious diseases, while living above ground was correlated with the propensity for other contagious diseases. Hence, what became important was a constellation of risk factors used to predict and eventually contain contagious disease outbreaks.

Epidemiological medicine precipitated a crisis in nineteenth-century clinical medicine (Petersen and Lupton 27–28). By the mid-twentieth century, professional case-by-case representations and analyses of disease presentations had ceded in importance to risk profiles for diseases based on surveillance and analysis of population aggregates. The creation of risk factors and profiles stimulated further surveillance aimed at prevention. Prevention technologies eventually expanded beyond the containment of disease vectors through hygiene to encompass more general strategies of health surveillance and promotion (e.g., regular health screenings, nutrition education, and exercise promotion) as medicine sought to prevent newly discovered prevalent conditions and chronic disease conditions (e.g., heart disease and type II diabetes).

Mid- to late twentieth-century epidemiological medicine was both prompted by, and contributed to, social welfare concerns about justice and national health. As explained in chapter 1, epidemiological medicine helped disseminate social welfare logics through its representational strategies and technologies of prevention, fostering popular support for more proactive government surveillance over, and promotion of, children's health and for government regulation of environmental hazards. Epidemiology therefore played an important role in shifting the burden of identifying and managing lead poisoning away from individuals toward social welfare institutions (e.g., NIH) and legislative bodies.

However, a convergence of factors in the late twentieth century led to a shift away from the use of social welfare logics to manage epidemiological risks. One important factor that helps explain the shift was the emergence of neoliberal policy orientations and problem-solution frames beginning in the 1980s. Neoliberal policy orientations toward managing health risks focus on using epidemiological surveillance to target at-risk individuals while then individualizing the responsibility for risk management. Diminishing support for public health apparatuses, particularly those dedicated to environmental health, such as the EPA, has in part prompted this shift. In addition, development of commercial health-screening services and products encourages logics that individualize risk. Personal risk management through consumption of commercial products and individualized technologies of the self thus assume special import in neoliberal regimes of health knowledge and preventive practices.

As will be explained later in this chapter, neoliberal value orientations and logics of government have captured and transformed approaches to dealing with lead poisoning. First, the valorization of market self-regulation and the politicization of science under the George W. Bush administration dampened EPA efforts to reduce lead emissions and to enforce cleanup of lead-contaminated urban areas. Second, recent media publicity about lead poisoning marginalizes poor populations, who bear the disproportionate burden of lead poisoning, in favor of accounts of lead in toys purchased by middle-class consumers. Efforts to educate consumers about the potential dangers of lead in toys have, in the popular press at least, replaced efforts to educate populations about the dangers of old paint, despite epidemiological evidence that the latter continues to pose the most pervasive and significant risk for undue lead exposure. When poor populations do appear in the media as victims of lead poisoning, recent reporting (e.g., Nevin 315–36) has tended to emphasize the apparent link between lead exposure and violent crime.

Explanatory accounts of violence that emphasize biological causation circulate widely and are slowly eclipsing social-environmental accounts. The biologization of crime and behavior more generally reflects new trends in epidemiological medicine as well as new technologies for representing biological bodies. In the 1960s and 1970s, epidemiological accounts cast individuals who had been exposed to lead as "victims" who "suffered" cognitive and emotional injuries caused by the *social crime* of lead poisoning. In contrast, recent epidemiological accounts of lead exposure have recast victims as potential perpetrators of criminal activity (e.g., Wright et al. 101). Popular and expert receptivity to this reformulation has been encouraged by the more general tendency for individual differences to be explained using biogenetic formulations. The biologization of childhood differences of behavior and intellect have important implications for the biopolitical sorting and treatment of children, as risk is located in biological substrates such as genes and neurology.

In 1953 James D. Watson and Francis Crick claimed success over at least six decades of efforts to identify the locus of heredity. Their model of DNA and explanation of its role in dictating the composition and synthesis of proteins transformed medical understandings of disease, ushering in a new age of medicine and a new social ontology of the subject (see Gottweis 53). The new social ontology of the subject articulates individual differences in behavior, cognition, and social success in terms of inborn biological predispositions coded genetically and materialized neurologically (see Nadesan *Govermentality* 138–72; Rose *Politics* 9–224). Socially problematic childhood disorders such as ADHD were captured by this new biogenetic ontology despite popular public sentiment and epidemiological evidence that biological environmental hazards such as lead play a role in causing the disorder. The new biogenetic ontology fundamentally diverges from the popular and epidemiological frameworks in that the former emphasizes the role of agentive genes in shaping behavior and cognition over environmental mediations.

ADHD thus offers a case example to explore how inborn, biogenetic explanations of children's behavior have gained academic and popular currency, shaping epidemiological understandings and treatment protocols. This section on children's health risks explores how inborn biological explanations of childhood deviance achieved prominence even in the context of environmental activism. This discussion contrasts with the displacement of inborn explanations in early to mid-twentieth century historical accounts of lead poisoning.

LEAD POISONING IN EPIDEMIOLOGICAL MEDICINE AND BIOPOLITICAL RISK MANAGEMENT

The story of the U.S. government's response to lead poisoning in children illustrates the limits of social welfare logics and technologies of government in the first half of the twentieth century. It illustrates how entrenched attitudes about heritable deficiencies limited the public's willingness to accept epidemiological evidence that lead was an environmental culprit capable of lessening intelligence as it slowly poisoned poor childhood populations living in dilapidated housing. It illustrates public unwillingness to regulate market authorities when childhood victims were members of marginalized groups. And it illustrates how epidemiological methods were part of a social matrix of political activism in the 1960s and 1970s that has now ceded to a very different kind of political nexus.

Lead Poisoning and Childhood Risk: History

Poisoning by lead was known to affect industrial workers in the nineteenth century, but was believed to be limited to workers engaged in particular activities and particular industries. The first recorded instances of children's exposure to lead poisoning in the home occurred in the nineteenth century, although these cases were rare and occurred through lead poisoning of food (English 7–13). In 1904 ten children died in Queensland, Australia, from nibbling lead paint off their porch's railings (Fee 575). This incident is widely cited as among the first-recorded instances of childhood lead poisoning from paint.

Researchers published reports of lead poisoning of children in the 1920s and 1930s using case-by-case studies to document the adverse effects of lead paint poisoning on children's health (Fee 576). Documented effects tended to be very overt and included symptoms such as comas and convulsions (573). More subtle symptoms reported included a change in disposition, abdominal pain, and anemia (575). These symptoms were all believed to derive from pica, the abnormal ingestion of nonfood items, particularly paint. Pica was believed to occur most frequently in lower-class populations, particularly African Americans (Washington 36). Rather than attributing pica to the probable medical cause of severe malnutrition, many medical authorities construed it as proof of individual and class degeneration.

Since public opinion held that only degenerate or deficient individuals were inclined to pica, political authorities saw little reason for regulating leaded paint. Leaded paint was considered to be superior to alternative paints, and the paint industry actively resisted any efforts to curb lead content. Children were even incorporated in lead paint advertising to imply the product's safety to children. The invention of leaded gasoline in the 1920s did raise some concern about potential public health effects. Ultimately, however, economic incentives outweighed health concerns (Rosner and Markowitz 344–52).

Despite the lack of political concern, a small group of medical doctors persevered in studying lead's effects on children. By the 1940s research published on lead paint poisoning in children had begun to develop a clinical picture of lead's internal effects on young bodies. For instance, in 1942 Paul Reznikoff used case studies to document that lead absorption by infants and young children creates deposits on the bone, concentrated on where the bone grows most rapidly (1125). The resulting "dense area" could be detected using X-rays. This internal examination of the body offered another strategy for diagnosing lead poisoning in young children (1125).

Clinical doctors' efforts to demonstrate lead's harmful effects on children were challenged by the research of Robert Kehoe, whose research on lead was supported by the lead industry (Berney 7). Barbara Berney chronicles how Robert Kehoe combated lead legislation in the United States from the late 1920s through the 1940s (7). Kehoe argued that lead occurred naturally in human tissues and excreta and that the body did not store lead; therefore, there was "no necessary relation between lead absorption and lead intoxication—no necessary connection between lead concentration in feces, urine, or tissues and lead poisoning" (cited in Berney 7). Kehoe's arguments were widely accepted until medical researchers began undermining his claims with demographic and epidemiological evidence in the 1950s.

Kehoe's arguments established the terrain upon which the adversaries battled. Accordingly, in order to prove that lead adversely affected children, researchers had to provide evidence for the following argumentative planks:

- Lead in the environment was a result of human use of lead in industry.
- Lead accumulated in the human body in proportion to the amount of lead found in the environment.
- Lead was absorbed by the body from the environment.
- Such absorption, measured in feces, urine, blood, and other tissues, was an indication of exposure and poisoning. (Berney 7–8)

The data generated from clinical case-by-case analyses provided insufficient evidence for supporting these claims' relevance for the entire childhood population. Therefore, it was not until the late 1980s that epidemiologists were able to provide indisputable evidence that low levels of lead absorption in the absence of pica cause bodily and neurological harm in children.

Efforts to build a case supporting lead's adverse effects can be traced to the proactive approach to health screening adopted by the city of Baltimore. In 1935 the city introduced a free diagnostic test for lead exposure. The test, used on people who were suspected of having lead poisoning, employed a new chemical method for detecting lead in blood called the "dithizone technique" (Fee 580). The data from Baltimore, the only city in the United States with a central diagnostic and reporting mechanism for lead poisoning, revealed lead poisoning as a frequent problem for children, particularly for those living in the city's dilapidated areas (Fee 581). Twenty-four percent of the 202 reports of childhood deaths from lead poisoning sent to the U.S. Bureau of the Census between 1931 and 1940were from Baltimore, due to its enhanced reporting mechanisms (Fee 583).

Research by Randolph Byers, a medical doctor at Harvard Medical School, and Elizabeth Lord, a psychology PhD, also played an important role in efforts to accumulate the necessary evidence. Published in 1943, their collaborative analysis of twenty case study histories demonstrated how non-fatal lead poisoning adversely impacted children's intellectual and emotional progress at school (Byers and Lord 479). At the time, blood levels of 70–80 µg/dL were commonly perceived to constitute poisoning, while levels of 50–60 µg/dL were regarded as normal (Berney 9). Byers and Lord note in their study that blood levels as low as 40 µg/dL could cause poisoning and that lead accumulated in patients' bodies and was retained after exposure stopped (Byers and Lord 475).

Bellinger and Bellinger observe that Byers and Lord's study was significant for helping transform the prevailing patient-oriented model of disease (855). The patient-oriented case-by-case model regarded poisoning to have occurred only when the child displayed specific and overt corporeal signs and symptoms of severe poisoning. This model presumed that children would recovery fully if they did not suffer from encephalopathy (855). Byers and Lords' research challenged both assumptions. Some of the children followed in their approach were admitted to the hospital for relatively mild signs of poisoning such as weakness, limping, irritability, and vomiting and did not exhibit the full range of symptoms of blood poisoning (476, 493). Yet after following up on these cases, Byers and Lord found that the children subsequently experienced marked academic and social problems. Intelligence tests administered by the researchers revealed relatively low scores ranging from 67 to109; children with higher measured scores tended to exhibit impulsive behavior (477). Byers and Lord's longitudinal case study approach revealed previously unrecognized long-term risks to children posed by exposure to lead in the absence of overt poisoning, or plumbism.

Time magazine ran an article popularizing Byers and Lord's findings. Titled "Paint Eaters," the article informs parents, "If your child is slow with building blocks, but quick on tantrums, he may be a lead eater." "One consequence of the plumbic passion in children may be stupidity," the article warns. This warning expanded the public's perception of the threat posed by lead exposure: lead didn't simply kill, it could also impair. Despite *Time*'s warning, the public tended to regard lead poisoning as a problem restricted to poor children and children who engaged in the purportedly deviant practice of pica. Retrospective analyses of case studies of lead poisoning pointed to deteriorating living conditions as the primary factor affecting exposure. The tendency for cases to be reported by place of poisoning rather than age after 1951 reinforced the idea that lead poisoning was a disease of poverty (Warren 16).

The idea that lead poisoning was a disease of poverty shaped attitudes toward its prevention. Public sentiment did not generally support eliminating lead from paint, nor did the public support government-enforced cleanup of the dilapidated housing implicated in lead exposure. Instead, the poor were regarded as indirectly responsible for their poisoning through their careless housecleaning, negligent parenting, and deviant children.

In the late 1940s, Baltimore sent public health nurses to poor city neighborhoods to educate mothers of young children about the dangers of lead paint (Fee 586). Mothers were interpellated as responsible for preventing lead poisoning by repainting their homes and by carefully monitoring their children. These efforts to make mothers responsible for ensuring their children's health through home hygiene extended decades of advice to mothers. For instance, in 1904 Adelaide Nutting of the Johns Hopkins Hospital published an article titled "The Home and Its Relation to the Prevention of Disease," arguing that proper home hygiene by mothers, including cleanliness from dust and plenty of ventilation and sunlight, prevents a wide array of diseases, including tuberculosis. Home hygiene, proper feeding of infants, and moral guidance together constituted the crux of responsible motherhood from the late nineteenth century onward through the mid-twentieth century. Consequently, the unintended effect of Baltimore's efforts to educate poor mothers about the dangers of leaded paint was that these mothers were responsibilized to prevent poisonings, despite the ubiquity of this danger in their home environment.

Poor children were also cast as responsible for paint poisonings. Popular media reports often emphasized the purportedly perverted appetites of paint victims. For example, a 1957 article in the *Saturday Evening Post* titled "Help for the Poisoned Child" includes an illustration of a poisoned African American boy being treated by a white medical staff. The article's text provides a stereotypical representation of "paint victims":

> Most paint victims are two- to three-year-olds and even younger crawlers at teething age. They nibble paint from toys, from crib railing, window sills; from scabrous walls in old tenements. The doctors call the paint eaters' ailment "lead intoxication," or "pica," a term defined as, "depraved or perverted appetite or craving for unnatural foods, such as chalk, paint, clay." (Berger 76–77)

The article reads ambiguously as to whether the "depraved or perverted appetite" derives from, or causes, the lead consumption. This representational framework no doubt contributed to a public unwillingness to regulate lead in paint because it indirectly cast children as responsible for their poisoning. In 1955 a voluntary standard for limiting indoor lead paint to 1 percent was passed, yet outdoor paint continued to have high lead levels (Medley 69).

In 1958 Baltimore initiated the first large-scale screening for lead paint in order to evaluate the prevalence of lead paint in Baltimore homes (Berney 8). Seventy percent of the 667 dwellings tested had lead in excess of 1 percent (8). Baltimore officials tabulated their lead-poisoning statistics for children on the basis of race rather than income (Fee 584). Black children's mortality from lead poisoning was found to be five times higher than white children's (584). The researchers were aware that black city residents lived in more dilapidated housing and that poor housing was the independent variable affecting mortality statistics. Still, rather than battling slum landlords, city health officials emphasized parent education and supervision, therefore reinforcing the idea that poor mothers' negligence was responsible for paint poisonings. The inner-city black child emerged as the iconic victim of lead poisoning in popular media representations. This tragic figure garnered some public sympathy but few policy protections.

In 1959 Randolph Byers suggested that pica alone might not explain paint-based lead poisoning in children. He observed that normal mouthing behavior of children in environments with high levels of lead paint could cause poisoning, even when paint was intact (Berney 9). Byers's observations would eventually help destigmatize lead poisoning as a disease of poverty and poor mothering. Epidemiological evidence played an important role in supporting Byers's supposition.

Declaration of the War on Poverty in 1964 made more money available for cities across the nation to implement citywide screening programs for childhood lead exposure (Berney 11). Data collected in the 1960s showed high incidences of elevated blood levels of lead; 25 to 45 percent of one- to six-year-olds living in high-risk areas had blood lead levels exceeding 40 μg/dL, then considered the upper limit of normal (Berney 13). More data allowed researchers to detect subtler effects. Each time screening guidelines were revised, researchers initiated new studies to determine whether the new level used to define "normal" was actually safe for children (Bellinger and Bellinger 253). Julian Chisholm's work in the mid-1960s suggested that children were more vulnerable than adults to lead's effects and that levels between 40 and 80 μg/dL might hold toxicological significance for children (English 147). In 1966 Harriet Hardy encouraged researchers to study infant exposure to lead, suggesting that infants might be particularly vulnerable to low levels of lead exposure (cited in Corn 107).

Field epidemiology and case findings reported in the 1950s and 1960s mobilized late 1960s- and 1970s-era activism around lead poisoning. Young epidemiologists in the 1960s and 1970s seized on lead poisoning as an issue that would bridge medical science and social activism: as Berney observes, "Lead poisoning could be used to tie the emerging environmental movement (and the reemerging consciousness of the environment in public health) to civil rights issues" (11). The black child poisoned by lead was increasingly cast as a victim of economic marginalization, environmental harm and injustice, and cultural stereotyping. Moreover, social activists could use growing epidemiological evidence of widespread lead poisoning among black children to challenge racist popular sentiments that held them to be cognitively inferior to their white counterparts. Lead provided an environmental explanation for gaps in the measured IQ of black and white children, deflecting both hereditary notions of inferiority and culturalist accounts of inferior parenting among poor black families. Activists used the environmental-biological frame of lead poisoning to combat the moral taint of poverty and racism.

Managing Risk: Epidemiology, Environmentalism, and Social Welfare Biopolitics

Broader public receptivity to bioenvironmental explanations was shaped by the growing environmentalist movement. Tracing the roots of the contemporary environmental movement is complicated, but popular awareness of the role human contaminants play in adversely affecting animal and human populations can be attributed to Rachel Carson's *Silent Spring* (1962) and to growing concerns about air pollution in major cities. *Silent Spring* explores the effects of DDT on natural environments and human populations. DDT was made available for civilian use in 1945 and was widely sprayed on agricultural crops. Carson's book was serialized by the *New Yorker*, beginning in June 1962, making it available to a wide audience and catalyzing concern about pesticides (National Resource Defense Council). Simultaneously, widespread concern was growing about air pollution. Although concerns about smog are hardly new, in the early 1960s articles began appearing in major newspapers providing scientific evidence of smog's adverse effects on health. The *Los Angeles Times* ran quite a few of these articles, as illustrated by a 1961 article detailing the link between smog and lung cancer (Nelson I1) and another article describing a study in Los Angeles examining a possible link between blood lead levels and smog (Nelson B3). In 1963 Congress passed the first federal Clean Air Act, which allocated $95 million to the U.S. Health and Education and Welfare Department to study and support research on air pollution and control. This act funded more smog research which, in concert with Carson's book, helped mobilize public concern about environmental risk.

Growing environmental concerns primed public receptivity to the results of the longitudinal data on lead poisoning and lead effects collected in the 1960s and 1970s. Drawing upon accumulating epidemiological evidence, public health authorities in the late 1960s and 1970s considered the possibility that lead exposure from a variety of sources over time could accumulate in the body, causing long-term health effects (Corn 104–107). Sources of lead exposure broadened to include gasoline and aerosols in the air from smelters (Berney 20–21; Corn 106). In additional, longitudinal epidemiological evidence collected in the 1960s and 1970s pointed to relatively subtle, but measurable, long-term cognitive and behavioral effects from subclinical lead poisoning (Bellinger and Bellinger 855). The idea of "undue lead absorption" was created and disseminated as a way of talking about toxicological effects in the absence of overt, clinical symptoms of lead poisoning (i.e., plumbism) (English 151).

In 1970 the U.S. surgeon general made lead poisoning an official health concern, orchestrating a shift in focus away from case findings and treatment of overt lead poisoning to prevention through mass screening and avoidance of undue exposure (Berney 14). The Centers for Disease Control (CDC) funded screening for approximately 4 million children from 1972 to 1981 using a newly developed finger-stick technology (Berney 14). Data collected from these screenings supported the argument that exposure could occur through everyday activities and was not limited to pica (i.e., oral consumption) (15). Simultaneously, the threshold for adverse effects from lead exposure for children was pushed even lower as retrospective studies linked mental retardation and learning disabilities to lead exposure in children whose blood tested positively but whom displayed no overt features of poisoning (16). The U.S. surgeon general reduced "undue exposure" of lead to 40 μg/dL (Berney 14).

Epidemiological data collected and analyzed in the 1970s called into question old stereotypes about childhood lead victims as the evidence gradually documented that all children, even those from affluent families, were vulnerable to subclinical lead poisoning. An article in the *American Journal of Nursing* in 1972 titled "Lead Poisoning: Silent Epidemic and Social Crime" reports that "for every child treated annually for lead poisoning, 25 children are injured by lead but receive no treatment. Since 12,000 are treated annually, this suggests that nearly 300,000 children go untreated each year"

(Reed 2181). The article cites pica as the mechanism for lead poisoning but claims that "studies have indicated that up to 50 percent of both middle-class and poor children demonstrate pica" (2182). The article provides two illustrations, one of a white child mouthing a windowsill and the other of an African American child eating peeling paint from an exterior wall. It concludes, "Lead poisoning is more a disease of the environment than of family neglect" (2184). While the article's language encourages the reader to regard lead poisoning as affecting all children, the imagery and conclusion reinforce the historical supposition that lead poisoning is a disease of poverty, disproportionately affecting inner-city residents. Yet, unlike earlier representations, this article shifts responsibility from the victims of lead poisoning to the broader society. Accordingly, the conclusion reads, "The disease is the product of social and political conditions," and encourages nurses to perhaps "become directly involved in the political process" to "promote health and housing as high community priorities" (2184).

Lead poisoning was thus transformed from an accidental occurrence afflicting suspect children to a "social crime." Undue lead exposure was articulated as a social justice issue in left-leaning publications such as the *Nation*. A 1978 article titled "Childhood's Hidden Epidemic" (Huebner 242) quotes a health department staff member reporting on a Los Angeles community: "One must...consider the additional burden on these communities in their struggle for social and economic equality when, as a result of elevated blood levels, many of their youth cannot compete favorably in school or on the job" (cited on 244).

Accumulating longitudinal data about the severity and scope of lead-poisoning effects at low levels of exposure was used to mobilize popular support for regulating lead. In 1971 the Lead-Based Paint Poisoning Prevention Act was passed by Congress, which prohibited use of lead paint on toys, furniture, and food utensils and established guidelines for paint used on residential buildings or constructed with federal funds (Medley 70). In 1976 the Toxic Substances Control Act (TSCA) passed, enabling the EPA (established in 1970) to control chemicals known to pose unreasonable risks to human or environmental health. Lead was entirely banned from paint in 1978, and catalytic converters were introduced and lead was phased out of gasoline beginning in 1975. Exposure to lead was thus legitimized as a broad-scale concern necessitating market regulation.

However, the expansion of lead's risk created confusion about the very definition of lead poisoning. Were children with elevated blood levels but no clinical symptoms really poisoned? Researchers in the 1970s debated whether subclinical effects such as school problems and failure could be explained by elevated blood levels or confounding factors (Berney 17). Some critics of epidemiological work on lead poisoning insisted that children with neurobehavioral problems such as low intelligence were likely to engage in pica, thereby confounding epidemiological conclusions that lead exposure caused the problems (17). Other researchers emphasized the mediating impact of diet and time spent in the home on lead absorption, implicating parental responsibility in moderating lead's harm to children's bodies. Old notions would not die that poor children were less closely supervised and were more likely to engage in deviant and developmentally inappropriate practices such as pica.

It was not until 1982 that researchers were able to use longitudinal epidemiological evidence to document unequivocally that undue lead exposure could "impede children's overall developmental progress" in the absence of medical symptoms of lead poisoning, including lead encephalopathy, palsy, seizures, and so on (Medley 63). A study published by Needleman, Gunnuoe, et al. in 1979 of low-level lead exposure in Boston schoolchildren's baby teeth concludes that children with high concentrations of lead performed significantly poorer than their low-lead counterparts on the Wechsler Intelligence Scale for Children (Revised) on three measures of auditory and speech processing and on a measure of attention (689–93). The lead industry responded by attacking Needleman's work, questioning his integrity (Rosner and Markowitz 330). Yet the lead industry's efforts to discredit his work were eventually overwhelmed by the vast amount of research, including animal studies, establishing the diverse biological and behavioral effects of lead (Berney 18).

It had taken years to collect the longitudinal evidence to establish conclusively that undue lead exposure caused adverse health and cognitive effects in all populations, irrespective of economic status. During the 1970s, a primary decade for this type of research, children's overall exposure to lead slowly declined as lead was banned from paint and from gasoline. Consequently, the threat to middle-class children declined as middle-class families moved into suburban homes built after lead was banned entirely from paint in 1978. Efforts from 1976 to 1980 by the National Center for Health Statistics to collect data on blood levels for the entire U.S. population using stratified random sampling produced a median result of 13 µg/dL (Berney 21). Regulation was working. Lead concentrations in outdoor air declined about 96 percent from 1980 to 2005 (Berney 21). The effect was that lead's relevance as a public-health concern for middle-class families declined at roughly the same time as indisputable evidence documented that all children were vulnerable to lead, even at low levels of exposure.

Middle-class parents were made aware of lead as a risk to their children's cognitive development but were able to reduce exposure through "lifestyle" choices, including where to live and how to avoid exposure through parenting practices. Unfortunately, working-class populations had fewer choices and opportunities. On average, black children aged six to thirty-six months were six times more likely to have blood lead levels of 30–39 µg/dL and eight times more likely to have over 40 µg/dL blood lead levels, as compared to white children in the period 1976–1980 (Nevin 326). Although undue lead absorption affected all children, the poor were disproportionately impacted.

As illustrated by an article published in October 1980 in the *Saturday Evening Post* titled "Suffer Little Children," lead poisoning lingered in the public imagination as primarily a disease of poverty. The article cautions, "Rich or poor, a child living in any home built before World War II can be the victim of lead poisoning," but six of its eight illustrations represent deteriorated housing and urban squalor. One illustration depicts a lead storage battery in a nondescript fireplace and another depicts a child playing in the snow with the warning that lead can accumulate in the ice of "heavily trafficked areas" (Frazier 73, 76). By implication, the article suggests that readers can reduce their exposure by careful selection of living conditions. Additionally, the article notes that "nutritional deficiency may also facilitate lead poisoning and increase its power to inflict damage," presenting yet another step for parents to pursue to avoid undue lead exposure (76). The most interesting aspect of this article is its author's conclusion:

> It is my contention that, although it is yet unprovable, at least a portion of urban apathy, violence, vandalism and the decreased performance of school children could be a result of increasing levels of lead in the urban environment. It is also my contention that it is past time that we find out if this is so. (77)

By linking lead exposure to urban "apathy, violence, vandalism," the author reinforces the decades-old idea that lead poisoning was a problem specific to poor, inner-city populations who failed to conform to "normal" standards of conduct; however, in an unusual twist, the author identifies lead exposure as the potential source of deviance rather than construing it as a by-product of deviant behavior.

Despite the growing expression of medical concern about lead exposure, inadequate public support prevailed for legislatively enforcing cleanup of deteriorating paint in rental properties. Additionally, little support could be garnered for tighter regulation of lead in consumer products beyond paint and gasoline. Members of the growing suburban middle class simply felt that their children were largely safe from the perils of undue lead exposure. Suburban physicians bolstered this position by their failure to recognize symptoms of undue lead absorption in middle-class populations (Medley 63–64). Sara Medley therefore describes lead poisoning as a "paradox in modern industrial societies" because of its "neglect" by medical professionals and the public despite clear evidence of its harms.

Activists who remained concerned about lead poisoning in the 1980s tried to amplify the public perceptions of the scope of the problem. Lead was found to pervade both built and natural environments. Although concentrated in urban slums, lead in the environment was found to be ubiquitous, ranging from Greenland's remote polar ice caps to rural areas (Medley 65). Lead was found in everyday household products ranging from newsprint to toothpaste (67). Research suggested that prolonged exposure to lead from multiple sources could be more detrimental than acute, severe exposure (68).

Research in the 1980s documented that detrimental effects could be found at ever-lower levels of exposure. H. L. Needleman's research during this time frame used longitudinal prospective studies that measured blood levels from birth (from cord blood), tracked children across time, and controlled for confounding factors by collecting data from diverse demographic groups. Effects were documented below 10 µg/dL (Berney 26). In 1985 the CDC lowered the level of concern to 25 µg/dL (Berney 28). By 1991 the level of concern had been lowered to 10 µg/dL (Bellinger and Bellinger). By 2003 government data were interpreted to indicate that "the threshold for harmful effects of lead remains unknown" (Meyer et al. 1). Childhood statistics collected by the U.S. federal government and released online in 2007 indicate that a child with a 10 µg/dL blood level experiences an average decrease in IQ of six points (Federal). In 2001–2004 approximately 1 percent of children aged one to five years exhibited blood levels in excess or equal to 10 µg/dL, a decline from approximately 88 percent of children in 1976–1980. Despite this decline, approximately 17 percent of black non-Hispanic children and 4 percent of Mexican American children had blood levels at or above 5 µg/dL in 2001–2004. Lead may be the most toxicological hazard for U.S. children, and lead from paint remains the most common source of lead exposure and toxicity (Fee, 573).

At the very beginning of the twenty-first century, most middle-class Americans living in suburbia were oblivious to lead poisoning until a barrage of media reports heightened public concern by focusing on two types of risk. First, statistical correlations with criminal behavior elevated the perceived social import of lead poisoning for the general population; second, a rash of reports about lead in toys imported from China captured the attention of middle-class parents fearful of neurological damage to their children. These will be discussed in the following section.

Lead-Poisoning Scares and Neoliberal Biopolitics

Lead captured national interest in the early twenty-first century when it was found in many products used by children. In 2007 lead was reported in the media to be found in polyvinyl chloride (PVC) used to line lunch boxes and children's vinyl bibs (Shin "Taking" D1). A variety of toys containing PVC were reported to contain traces of lead and other dangerous chemicals, including cadmium. Lead was found in toy jewelry and other trinkets imported from India and China and in the paint of toys manufactured in China by well-known U.S. toy producers, including Mattel. The U.S. Consumer Product Safety Commission (CPSC) issued recall notices for 6.7 million pieces of children's jewelry in 2007 because of dangerous levels of lead (Merle "Recalls of Toys" D1).

These reports began in response to media publicity surrounding lead poisoning in a few cases where children had swallowed small toys and jewelry containing lead. Accounts stimulated widespread and often independently conducted testing of children's products for lead content (Story). None of the media reports evaluated for this chapter included evidence of increased average or median levels of childhood blood levels. Still, lead in toys was transformed from invisibility into a pressing national issue.

Heightened public anxiety about lead exposure reflects media attention but also points to deeper anxieties about middle-class children's abilities to excel in the "cognitive age." As explained in the last chapter, toys are no longer regarded as mere playthings, nonproductive occupiers of children's time, but are viewed as tools capable of enhancing children's "cognitive development." Toys are a kind of "ritual magic" seen as warding off the detrimental cognitive and emotional "risks" of institutional child care, television, and so on (Nelson-Rowe 117–31). The idea that toys might contain elements capable of reducing children's intellectual achievements violates the very principles of their active consumption.

The perceived risks of potentially dangerous toys also tapped into populist resentment against global competitors. Most toys today are manufactured in China. American middle-class workers typically express deep anxiety about the neoliberal economic flow that is perceived as threatening their ability to maintain a "middle-class" lifestyle as jobs are outsourced to the capable and cheaper workforces of India and China. Contaminated toys from China may tap into deeply held anxieties about national and individual competitiveness and may subconsciously have been viewed as a kind of sabotage.

An American public suspicious of government inefficiency and enamored with industry self-regulation was therefore shocked to discover that their most trusted manufacturers had inadvertently introduced poisoned goods into middle-class households. The public learned that although federal law bars lead in paint and dishes, it does not specifically bar lead used in manufacturing plastics and other materials. Moreover, the public learned that the CPSC lacked regulatory capacity because of decades of underfunding:

> The CPSC has suffered for three decades from slashes in budgets and staffing initiated during the antiregulatory fervor of the Reagan era. The agency of about 400 acknowledges its struggle to police more than 15,000 products in a global marketplace where 80% of toys sold in the U.S. come from China. (Trottman and Williamson A3)

Toy distributors and manufacturers publicly expressed favor for new federal regulations, believing they would be less stringent than those legislated by some states. Public safety advocates pressed for online data banks of consumer reports of product-safety problems online, but manufacturers lobbied hard against public access to these data banks (Trottman and Williamson A3). Online databases are consistent with neoliberal strategies of self-government by "informed consumers."

For the last twenty years, the U.S. government has facilitated market self-regulation and advocated consumer awareness and choice over government oversight and regulation. In August 2007 the U.S. Government Accounting Office (GAO) published a report comparing the lax U.S. regulatory framework for chemicals with a recently enacted European framework, REACH (Registration, Evaluation, Authorisation, and Restriction of Chemical Substances). The GAO report, titled "Chemical Regulation: Comparison of U.S. and Recently Enacted European Union Approaches to Protect against the Risks of Toxic Chemicals" (GAO-07–825), explains that under the current regulatory system in the United States, companies do not have to provide information on the health or environmental impact of chemicals unless specifically required by EPA ruling. Consequently, the EPA relies on voluntary programs for gathering information from chemical companies in order to evaluate and regulate chemicals under the provisions of TSCA. The GAO report found TSCA inadequate in comparison with REACH's reporting requirements. Moreover, as encapsulated in the GAO executive summary:

> TSCA places the burden of proof on EPA to demonstrate that a chemical poses a risk to human health or the environment before EPA can regulate its production or use, while REACH generally places a burden on chemical companies to ensure that chemicals do not pose such risks or that measures are identified for handling chemicals safely.

The GAO report's recommendation that the burden of risk be shifted to the chemical companies was not adopted by the George W. Bush administration.

Popular outrage over contaminated products imported from China, among other places (e.g., Mexico), did eventually lead to public support for a more stringent regulatory environment for children's toys. Under a lawsuit settlement with the Sierra Club, the EPA began in 2008 requiring manufacturers and importers to file health and safety reports when lead is detected in many children's products. The CPSC issued new lead standards in September 2008 but allowed companies to sell off existing products before the new standards took effect in February 2009 (Shin "Tighter Lead" A9).

Ironically, all the attention afforded lead in children's products did not coincide with renewed media attention to lead paint, which continues to be the primary source of lead exposure for children with elevated lead levels. No national discussion of removing lead from dilapidated housing has ensued, although a CDC report estimated that in 2000, 25.8 million housing units continued to have lead paint (Meyer et al.). The principal sources of lead exposure for U.S. children continue to be house dust contaminated by leaded paint and industrial and auto emissions. Non-Hispanic black children, Hispanic children, and poor children continue to be at greatest epidemiological risk for undue lead exposure (Meyer et al.).

Expert epidemiological accounts of the greatest sources of risk for lead exposure have little resonance with media reports and congressional debate, both of which continue to focus on lead in consumer products. While biological risks are posed by consumer products, their epidemiological significance is far less than the risks posed by paint in older buildings and airborne lead exposure. In effect, poor children are not a regarded in the public imagination as "at risk" unless they inadvertently swallow a toy that could also pose risks for middle-class children.

Interestingly, the marginalization in the public imagination of poor children as "at risk" for lead poisoning roughly coincided with new ideas about lead-poisoned children as "risky." Longitudinal studies finding correlations between lead exposure and criminal activity began to circulate in the late 1990s. These accounts inadvertently represented lead-poisoned children as posing public safety risks. Herbert Needleman's work again played a significant role in stimulating public concern about the effects of lead, as he and his research associates linked bone lead levels and delinquent behaviors (e.g., Needleman, Riess, et al. 363–69). By 2007 meta-analysis linked preschool lead exposure to international crime trends (Nevin 315–36), and longitudinal research established a firm connection between prenatal and early childhood exposure to lead and subsequent criminal behavior (J. P. Wright et al. 101).

Efforts to link lead exposure to delinquency and criminality no doubt were spurred by multiple motivations. As mentioned previously, social and public health activists have had little success in enforcing lead paint removal from the dilapidated housing in which many poor populations continue to live. Poor children continue to be poisoned by lead paint in older buildings, and many continue to have blood levels believed to pose long-term risks to their cognitive and emotional health. Yet there are no national regulations requiring elimination of lead paint in preexisting buildings in the absence of documented poisonings. By linking lead exposure to public safety, activists may have found a strategy for building greater support for combating lead poisoning.

But this move to link lead exposure to criminality and delinquency has implications for societal formulations of risk. Since the 1960s, the child at risk for lead poisoning was constructed in the public imagination as vulnerable and in need of protection. This protective attitude prevailed even when societal sentiment viewed children as contributing to their poisoning through abnormal behavior such as pica. And it prevailed even when mothers were held responsible for their children's poisoning because of their supposed lax oversight. This protective attitude is clearly evident in popularized accounts of lead poisoning as illustrated by the 1957 article on poison control centers in the *Saturday Evening Post*. Although this protective attitude ultimately was outweighed by property owners' economic interests and laissez-faire regulatory attitudes, the poisoned child and the child at risk elicited some public sympathy and activated a generation of scholars and epidemiologists.

But recent accounts linking criminal activity and violent crime to lead exposure may undermine public sympathy for the lead-burdened child, as images of childhood innocence are replaced by visions of impaired, dangerous criminals. Efforts to link lead to delinquency and criminality resonate publicly in part because of a growing biologization of crime and behavior. The move toward biological approaches to explaining skills, emotional states, and social outcomes has significant social importance for the biopolitical sorting of children for risk. The next section explores how biological accounts of children's behaviors, traits, and social outcomes are overshadowing and/or marginalizing environmental accounts.

FROM ENVIRONMENTAL VICTIMIZATION TO
HERITABLE RISK: THE BIOPOLITICS OF ADHD

As explained in chapter 3, biological accounts of delinquency, criminality, and mental illness prevailed throughout much of the second half of the nineteenth century. Case studies were used to document the inborn characteristics of dangerous individuals. Researchers then attempted to find related individuals who also exhibited the same problematic characteristics, inferring heritability from familiar frequency groupings. Late nineteenth- and early twentieth-century psychiatry strove to identify "stigmata of degeneration" of inborn "moral" character deficiencies in children, as illustrated in George Still's 1902 account in his series of articles titled "Some Abnormal Psychical Conditions in Children" (1081). Despite the pessimistic fatalism of authorities such as Still and Henry Goddard, child savers had successfully institutionalized programs that stressed prevention and treatment of delinquency by the 1930s. Popular dissemination of psychoanalysis in the period ranging from the 1930s through the 1950s contributed to the public's willingness to consider environmental explanations for childhood deviance. Publicity about Germany's atrocities during World War II also tempered some of the more virulent expressions of eugenic attitudes in the United States, although racist and ethnic biases persisted.

However, beginning in the late 1950s, biological and genetic explanations of human behavior and diversity began to appear more frequently in popular press accounts of scientific research. Evolutionary biology explored social behavior, creating sociobiology. Biologists searched for and "discovered" the purported elements of heredity. Psychiatry drew parallels between human emotional states and specific brain chemicals. Although disparate, these biological discourses all shared one common theme: human social differences are derived from inborn biological agents. Print journalism played a particularly important role in disseminating these new biological findings to the public. Popular media accounts offered the public new, biologically based ways for talking about social problems (e.g., drug addiction and juvenile delinquency) and for explaining differences in individual children's outcomes (e.g., cognitive test scores and school performance). These accounts tended to deemphasize the significance of exogenous, or environmentally based, biological factors such as pollutants like lead or nutritional factors in shaping cognition, skills, or affect, stressing instead inborn genetic and neurological characteristics. Pharmaceutical advertising that encouraged the public to view their bodies and minds physiologically helped condition popular receptivity to biological accounts of human emotions and individual differences. For instance, as will be illustrated in this section, early discussion in the popular press of what is now known as ADHD included reference to the seemingly miraculous drug that normalized middle-class children afflicted with "minimal brain dysfunction" (Clements and Peters 185–97).

My argument is that these new, inborn (i.e., endogenous) biological formulations today undermine public support for collectivized and preventive risk-management laws, policies, and institutions. Biologically determinist accounts of traits and behaviors discount the impact of social interventions. Although many people reject biological determinism, the cultural resources available for adults seeking to help at-risk or risky children privilege pharmaceutical solutions, which again imply that there exists a biological basis to the risky behavior targeted for treatment. Adults who look beyond medical and psychiatric expertise often confront a marketplace of self-help books, Web pages, and products, most of which stress parents' responsibility for saving their children through the consumption of particular commodities or services. The discussion that follows traces the ascendancy of biological thinking by exploring how research about heredity became more central in social discourse about children's capacities and measured traits. Scientific discourse about ADHD is used to demonstrate how inborn biogenetic explanations have marginalized environmental accounts of children's behavior and educational outcomes. ADHD is a particularly relevant example for studying the rise in genetic explanations because the disorder's symptoms are also characteristic features of childhood undue lead exposure (see Braun et al.).

ADHD Historically

The history of ADHD aptly illustrates how explanations for troubling childhood behaviors reflect both social values and dominant paradigms of understanding. The behaviors characteristic of what we today consider ADHD were articulated in the very early 1900s as inborn, biologically based phenomena but eventually were transfigured as socially derived symptoms of emotional maladjustment. Contemporary understandings of ADHD as a biological, heredity-based disorder were articulated in the early 1960s in the context of renewed social concerns about delinquency and crime and a resurgence of biological accounts of human behavior in both expert discourses and in the popular press. Although much has been written about the history and criticism of ADHD, this section will not rehearse this extensive and fascinating literature but rather will focus specifically on how ADHD was framed as an *inborn* biogenetic condition treatable by the medication Ritalin. The discussion also explores why this account may have appealed to everyday people seeking to meet societal expectations.

ADHD-type symptoms were observed by biopolitical experts prior to the labeling of the disorder. In 1897 Fletcher Beach of London describes a hyperkinetic boy "on the move since birth" because of outbursts of screaming and destructive violence (cited in Walk 764). Beach ascribes this type of childhood deviance to a "perversion of affective faculties" attributable to masturbation (765). In 1902 George Still attributes lack of moral control, or a "morbid defect of volition," to a wide array of biological diseases and conditions in Lecture I of his account of children's abnormal psychical conditions (1008). In Lecture II, Still expresses concern over the risk of contagion posed by these children:

> The pernicious influence which some of these morally defective children may exert on other children is appalling to think of. I have said very little about the sexual immorality which accompanies the other manifestations of morbid defect in some of these cases, but it is too important to be disregarded. (1167)

Still emphasizes the "paramount necessity for separating some of these morally-defective cases from other children," including the possibility of "confinement in special institutions" (1167). Although his account is tinged with moral condemnation, Still also raises the question of "how far these children are to be held responsible for their misdoings" given the (purported) biological bases of their behavior (1167). By medicalizing these children's "deficiencies," Still lessens their legal culpability; however, he also raises the risk of contagion. His approach is therefore infused with eugenic anxieties and a disciplinary problem-solution frame geared toward identification and enclosure.

The child guidance clinics established in the first decades of the twentieth century provided opportunities for psychiatric authorities to collect and analyze data from a wide range of children referred to the clinics for their delinquency or for subtle aberrations in behavior linked to "pre-delinquency" by the 1930s. Data-collection efforts were driven by the desire to prevent future deviancy by identifying those at risk. Accordingly, research findings reported in the 1930s mostly consisted of surveys of observations of normal and delinquent behavior patterns (e.g., see Childers 227–43).[2] Hyperactivity emerges in this literature as a frequently mentioned behavioral characteristic of children diagnosed with "conduct disorders" (see Harwood 50). Hyperactivity (also "restlessness") was seen as a sign of poor intelligence or conversely as a sign of above-normal intelligence (Blackman 54–66; Selling 92–93). Hyperactivity was seen also as a symptom of insecurity (Childers 227–43) and lack of sleep (Bridges 536). Hyperactivity was regarded as a symptom in these accounts of a general class of "conduct disorders" that placed the child in question at risk for delinquency.

Clinical efforts to help children with conduct disorders in the 1930s occasionally employed medications; however, their use does not mean that behavioral symptoms were regarded simply as surface manifestations (i.e., epiphenomena) of biologically based defects. Charles Bradley, who experimented with Benzedrine (generic *dl*-amphetamine) on children hospitalized for significant behavioral and school difficulties, observes in 1932 that "in discussing emotional responses it must be kept in mind that in many of these children the conduct disorder was essentially an external symptom of emotional maladjustment" (579). Bradley's text is agnostic about the source of the maladjustment, but his approach seems to fuse social and biological explanations by implying that the maladjustment impacts the central nervous system. This linkage is important for his account of the paradoxical sedating effect of the stimulant drug (582–83). Bradley ultimately cautions against using the drug without further clinical trials because of troubling side effects (583).

As evidenced by Bradley's account, biological psychiatry played a role in the 1930s-era child guidance clinics, but most clinical authorities drew upon social-psychological explanations as well. Adolph Meyer, who played an important role in founding the mental hygiene movement in the United States, adopted a psychobiosocial approach incorporating both biological psychiatry and psychoanalysis (Stone 153). Meyer significantly influenced Leo Kanner, who published *Child Psychiatry* in 1935. Kanner's account of childhood autism illustrates this psychobiosocial framework. Drawing upon Eugen Bleuler's (1857–1939) characterization of autistic aloneness in schizophrenics, Kanner describes autism as a heritable and biological condition shaped by parental psychodynamics, particularly by maternal affective behaviors (Nadesan *Constructing* 65–67).

Kanner's and Meyer's psychobiosocial accounts of childhood deviance contrast with the simple hereditary notions promulgated by figures such as H. S. Jennings, who published *The Biological Basis of Human Nature* in 1930. Jennings describes how individuals are concocted "as it were on diverse recipes; and the diverse recipes give different results...Some combinations of them [i.e., cells] give imperfect individuals, feebleminded, deformed monstrous. Others give normal individuals, others superior individuals" (2–3). Jennings's perspective was shared by medical authorities who feared that subtle deficiencies of intelligence in children engender moral deviance (Harwood 41). The idea of the psychopathic child was born from this idea of heritable, inborn deficiency and deviance.

As illustrated by these contrasting accounts, expert understandings of children's behavioral difficulties reflected a wide range of influences, including hereditary flaws, defects of birth, and psychobiological dynamics. However, despite differences in beliefs about the cause of deviance, most child guidance authorities in the 1920s and 1930s believed that a child's character was ultimately malleable. Even the psychopathic child was believed responsive to effective governance, as demonstrated by the proliferation of child guidance clinics in Europe and the United States. Psychiatric authorities of the time stressed the role of the home environment in tempering or exacerbating any inborn tendencies toward delinquency. For example, a 1932 article, "Children from Happy Homes Have Less Trouble in School," published in *Science News Letter*, reports the effects of marital discord on children: "Personal maladjustment following marital disharmony as noted in school may be marked by hyperactivity, inferiority, fearfulness, reverie, mental conflict, emotional instability, sensitivity, self-consciousness, seclusiveness" (352). Valerie Harwood points out that the organizing drive of the child guidance movement was to shape children's character through effective parenting, guided by expert authority, so as to prevent disorderly and deviant behavior (39).

Research on children declined during the World War II period in the 1940s. Research published during this period tends to promote social-psychological and psychoanalytical approaches to understanding children's emotional adjustment. The cultural preoccupation with children's intelligence characteristic of 1930s-era research faded, while personality and social relations grew in importance. Social explanations for children's behavioral subtleties achieved greater visibility. For example, in 1940 Ruth Pearson Koshuk writes in a monograph titled "Social Influences Affecting the Behavior of Young Children" of "serious attempts" within the last two decades to study the "influence of social factors, broadly defined, on the behavior of young children" (1). The child's relationship with his or her mother stands in this discourse as the most critical factor shaping psychopathology and deviance. Biological psychiatry is less evident in research citations, although Andrew Lakeoff's review of the history of ADHD finds one 1940s-era psychiatric account of "minimal brain damage" used to describe nonencephalitic patients who exhibited motor-impulse problems (Lakeoff 152). In addition, one summary of trends in diagnoses at a mental hygiene clinic from 1929 to 1947 reveals the "increasing conviction that hyperactive, impulsive behavior, like epilepsy, has a constitutional basis" in brain injury (Doering 87).

Psychoanalysis exerted significant influence over psychiatry in the 1950s, shaping expert explanations of troubling childhood symptoms and behaviors. A 1958 text titled *Mental Subnormality* describes "hyperkinesis" as a symptom of impulsiveness (318), represented as a sign of "emotional dissociation" (Masland, Sarason, and Gladwin 317). As illustrated by this account, hyperactivity remained at this time a symptom more of significant psychiatric disturbances rather than of a distinct disorder. The *Diagnostic and Statistical Manual of Mental Disorders I* (*DSM*), published in 1952, does not offer a distinct disorder capturing the behavioral and cognitive difficulties later grouped as ADD/ADHD, nor does it provide a category describing the antisocial behavior of children (DeGrandpre 38). This first version of the American Psychological Association's (APA's) nosology of psychiatric conditions is decidedly agnostic about the etiology of most of the forms of mental illness it describes (Transit 48), despite providing typologies of a vast array of psychopathic personality disturbances.

The decade of the 1950s is significant for the explosion of experimental research using pharmaceutical drugs to control children's behavioral difficulties. Newly created psychiatric drugs promised to manage distressing symptoms in children in the absence of clear understandings of their etiology. For instance, *Science New Series* reports in 1956 that "the behavioral disorders that commonly afflict mentally retarded children—ranging from destructiveness and breath holding to psychogenic vomiting and teeth grinding—have responded in many cases to treatment with tranquilizing drugs, particularly chlorpromazine" ("Tranquilizing Drugs" 259). The article notes that chlorpromazine is effectively used to treat "vomiting and abnormal appetite, restlessness, lip sucking, hyperactivity, anger, cruelty and aggression, negativism", and so on (259). This antipsychotic drug, synthesized at the end of 1950, was apparently used primarily on children designated as mentally retarded who exhibited hyperactivity symptomatically (Bair and Herold 363–64). As Lakeoff points out, the use of pharmaceutical drugs was unproblematically incorporated within agnostic and psychodynamic explanations of the source of symptoms targeted for medication (158).

Popular 1950s-era understandings of childhood deviance in the absence of more significant problems of psychosis or cognitive ability placed blame primarily upon the mother. For example, the *Saturday Evening Post* reports in 1956 that "when a boy or girl doesn't get passing grades, the parents are usually to blame" (Donatelli 51). Psychoanalytic accounts popularized by Winnicott in 1951 (e.g., "Transitional Objects and Transitional Phenomena") and Bettelheim in 1967 (e.g., *The Empty Fortress: Infantile Autism and the Birth of the Self*) attribute childhood misbehavior, neuroses, and even psychoses to maternal indifference or overprotectiveness. Biological psychiatry, overshadowed by these paradigms, awaited reinvention.

ADD/ADHD in the 1960s and 1970s: The Biologization of a Disorder

Biological psychiatry was in a sense reborn in the 1960s and 1970s with the advent of new technologies for understanding heredity (i.e., genes) and the brain, coupled with the rebirth of evolutionary psychology. Simultaneously, however, social-environmental approaches to explaining deviance in children's behavior and aptitude scores also grew in importance. A retrospective analysis of the research of the period beginning in the 1960s reveals a schism that occurred as (1) biological psychiatry and genetic science medicalized middle-class children's deviance with the development of new diagnostic categories such as ADD (later named ADHD) while (2) environmental science, epidemiology, and sociology claimed purview over poor children using lead poisoning, nutritional deficiencies, and the culture of poverty as explanatory frameworks for behavioral disturbances and lower aptitude test scores. By the dawn of the twenty-first century, biological psychiatry and genetics eclipse social-environmental accounts, embracing all children within their purview of authority.

At the end of the 1950s, studies began to emerge that identified hyperactivity as a distinct disorder unto itself (Freeman 7). "Hyperkinetic impulse" disorder was coined in 1957 by Laufer, Denhoff, and Solomons (38–49). The authors claim:

> A very common cause of children's behavior disorder disturbance is an entity described as the *hyperkinetic impulse disorder.* This is characterized by hyperactivity; short attention span and poor powers of concentration; irritability; impulsiveness; variability; and poor school work. The existence of this complex may lead to many psychological problems, due to the extremely irritating effect it has upon parents and teachers. (38)

The authors attribute the disorder to "dysfunction of the diencephalons" (38), causing the cerebral cortex to be exposed to "intense storms of stimuli from peripheral receptors" (38). The authors note, "The photo-Metrazol determination done in children showing this syndrome is significantly different from those without the syndrome" (38). Amphetamines are reported to increase the "photo-Metrazol threshold" in such children until maturational processes reverse the condition.

As illustrated by this account, articulation of hyperkinetic disorder as a distinct disorder was legitimized through the use of EEG brain-screening technology. The "photo-Metrazol" determination involved use of EEG technology to measure the brain's electrical activity using electrodes placed on the scalp. EEG topography developed in the 1950s promised to provide visual maps of EEG data. This technique was employed to reveal the (supposed) flawed neural-electrical interiority of otherwise normal-appearing children and adults (e.g., see Werry 9–16). Although EEG data can reliably indicate brain tumors and epilepsy, its capacity to represent psychiatric disorders has been subject to considerable debate. In the early 1960s, this debate was muted, and psychiatry welcomed the role of EEG data in proving empirically the neurological reality of subtler forms of mental aberration.

"Minimal brain dysfunction" was coined in 1962 as an explanatory catch-all for hyperkinetic disorders (Clements and Peters 185–97). In 1968 the *DSM II* decided to label the loosely understood and ambiguously identified emerging disorder of control "Hyperactive Reaction of Childhood," which evolved into "Attention Deficit Disorder" with the 1989 *DSM III* (DeGrandpre 38; Lakeoff 153). Leo Kanner's *Child Psychiatry* did not introduce the diagnosis until 1972 (Lakeoff 153). Once a symptom of some other more serious underlying pathology, hyperactivity had evolved by the early 1970s into a distinct disorder worthy of its own entry in the most revered nosologies of the time.

The public was introduced to the idea of this disorder in the late 1960s and early 1970s at the same time stories of middle-class children's school difficulties, behavioral problems, and juvenile delinquency began to appear more frequently in the popular press. A 1970 article in the *New York Times* titled "RX for Child's Learning Malady" describes a disorder of "minimal brain dysfunction" featuring a white, apparently middle-class child with his mini-skirted therapist (Reinhold 27). The article claims that "best estimates" hold that "5 to 20 percent of American children suffer from this disorder, making it a problem of epidemic proportion" (27). Luckily, the article explains, a new drug now available can treat its troubling symptoms. A 1972 article, "Experts Now Link a Learning Disorder to Delinquency," reports a lower frequency of the disorder but raises alarm by linking it not only to school failure, but also to criminality:

> Experts on learning disabilities…noted that the neurological disorder, which affects at least 3 percent of the nation's schoolchildren, is finally being

recognized as a major cause of school failure, emotional disturbance and even juvenile delinquency. (Brody 36)

The article identifies heredity as a main cause of the disorder and features a white child. This type of representation medicalizes white children's deviance by carving out a new biological syndrome capable of explaining educational and "moral" failures of white, middle-class children. The medicalization of deviance and educational failures deflect attention from the home environments and child-rearing capabilities of white middle-class parents within a framework that absolves parents of responsibility for their children's deficiencies.

Minimal brain dysfunction and ADD emerge within this period as medicalized labels useful for explaining middle-class children's school or behavioral difficulties (see Conrad and Schneider 156–61). The statement that ADD is a medicalized label does not affirm or deny the material reality of ADD/HD, nor does this claim purport that middle-class children alone received diagnoses of minimal brain dysfunction and ADD. Rather, the point is that historically specific concerns and problem-solution frames, including the resurgence of biological psychiatry, new (Cold War) concerns about middle-class children's intellectual capabilities, middle-class anxiety about youth rebellion, and the disabilities rights movement, converged to produce a new way of thinking about relatively subtle expressions of middle-class childhood deviance. Release of federal dollars to states to help educate children with disabilities in 1965 and 1968 encouraged heightened surveillance of middle-class children by schools and prompted research studies exploring the special educational needs of children exhibiting hyperactivity.

What is particularly interesting about the framing of minimal brain dysfunction and ADD is the discounting of environmental insults or home environments in producing hyperactivity by the 1980s. The child with hyperactivity was represented in the 1960s and 1970s as irrevocably "brain damaged." Birth trauma was suggested as causing hyperactivity, but hereditary studies using biometric methods increasingly claimed family members of children with hyperactivity suffered from an array of psychiatric disorders (Cantwell 414–17; Morrison and Stewart 189–95; Stewart and Morrison 209–12). The idea that subclinical lead exposure might cause the disorder was introduced in the early 1970s (see Needleman 47–54) but was discounted or regarded as highly speculative throughout that decade (see Freeman 13). The primary "environmental" focus, beginning in 1973, was food additives, which were quickly excluded by research authorities but still widely held suspect by the public (Kolata 516). ADD/ADHD was emerging in expert discourses as a biologically based, hereditary disorder despite popular suspicion that modern food additives or poor parenting were to blame. Yet the increasingly "hereditary" constitution of the disorder did not seem to cast aspersions upon middle-class parents, probably because the disorder was deemed "treatable" in the context of the prevalent, humanistic ethos of self-actualization that had captured and defined middle-class childhood.

Treatment tended to emphasize pharmaceuticals. Methylphenidate emerged as the drug treatment of choice in the 1980s (Werry et al. 292–312). By 1968 stimulants, including dextroamphetamine and methylphenidate, were studied clinically and began to be more widely prescribed by pediatricians (e.g., Millichap et al. 235–44; Weiss et al. 145–56). In 1970 critics began attacking the use of drugs to manage the disorder (Freeman 6). Rick Mayes and Adam Rafalovich claim that pharmaceutical companies consequently avoided funding research addressing stimulants' effects on hyperactive children (435–57); however, pharmaceutical companies were not averse to providing researchers complementary medication. For instance, the drugs for Millichap's 1968 study were provided by SmithKline and French Inter-American Corporation (235). Pharmaceutical companies also helped parents advocate for the inclusion of ADD within the 1975 Education for All Handicapped Children Act (Lakeoff 162). By the late 1980s, approximately 1 million children in the United States were taking the brand name form of methylphenidate, Ritalin, regularly to "control" their symptoms (DeGrandpre 18).

Critics located on the margins of popular discourse (e.g., "alternative" health proponents) and academic scholarship (e.g., medical sociology) questioned the disorder's reality. In 1975 Peter Schrag and Diane Divoky published *The Myth of the Hyperactive Child: And Other Means of Child Control.* This book contributed to a literature exploring and critiquing medicalization of childhood (e.g., Brancaccio 165–77; Conrad and Potter 559–82; DeGrandpre 1–249; see review by Miller and Leger 9–33). Critics of childhood medicalization and medication attacked Ritalin and the pharmaceutical management of childhood (Breggin 1–431). Other critics argued that ADD symptoms were "real" but were caused by diet, allergies, or environmental pollutants.

These critical debates largely disappeared from mainstream popular press accounts in the 1980s. ADHD had crystallized within the mainstream media as a medicalized disorder characterized by a set of homogenized clinical symptoms and treatment protocols. For instance, in 1985 *Parents Magazine* reports authoritatively in an article titled "The Truth about Hyperactivity" that ADHD has a biological basis and may be helped through a combination of educational assistance, therapy, and drug treatment (Rubins 110–12).

Many mothers' receptivity to inborn, biological accounts of ADHD in the 1970s and 1980s no doubt stemmed from their feelings of frustration and inadequacy and their failure to achieve their ideals of the "good mother" (Singh 1193–205). Inborn, biological accounts absolve mothers of direct responsibility and offer the expertise of medical authorities. Biological accounts often suggest clear treatment protocols in the form of easily administered medication, and ADD/ADHD was introduced in the press with a prepackaged pharmaceutical solution. Even critics acknowledge that Ritalin can at least temporarily increase concentration among nearly all who take it. Mothers desperate to enhance their children's success and school manageability, while responsibilized for their care, may therefore prefer biological accounts.

However, not all mothers accepted this inborn biological explanation, nor did all accept Ritalin as their treatment of choice for difficult-to-manage children. Toby Miller and Marie Clare Leger describe a "moral panic" surrounding Ritalin that seeped into popular media in the late 1980s (9, 25; see also Lakeoff 25), leading some mothers to adopt alternative therapies in lieu of pharmaceuticals (Malacrida 366–85). Alternative approaches adopted today range from diet to behavior-modification strategies, requiring intense disciplining of the child and his or her environment. The natural health movement has particular appeal for mothers trying to rid their child of distressing symptoms by purifying diets of unnatural chemical additives. Craig Thompson observes that for many, the natural health movement "stands as the post-modern apogee of the therapeutic worldview" (82) but points out that the movement's adherents find agency in its emphasis on consumer education and self-help care. Mothers may regard alternative health technologies as more empowering than those technologies employing mainstream medical expertise and prescription drugs because alternative technologies encourage maternal agency through research and intensive parenting protocols (see Malacrida 366–85).

The rise of the Internet as a widely used home-based information technology in the late 1990s facilitated widespread circulation of nonpharmaceutical treatment protocols for ADHD symptoms. The explosion of self-help literature on the Internet for troubling childhood symptoms points to the hyperresponsibilization of mothers for their children's health, but does not necessarily problematize biological understandings of childhood disorders. Internet-based self-help information sources usually draw upon biomedical models of disease processes to explain troubling childhood disorders such as ADHD and autism but are distinguishable from expert medical knowledge by an economy of "hope" (Rose *Politics* 135), which promises consumers the capacity to treat, alleviate, or cure troubling symptoms (Nadesan *Constructing* 166–67, 194–96). The economy of hope often derives from the idea that troubling childhood symptoms are caused by the child's inborn biological susceptibility to allergens, pollutants, or metabolic dysfunctions. In contrast to the medical model promulgated by medical experts working in clinics and hospitals, Internet formulations of biological susceptibility tend to stress the child's malleability and potential responsiveness to changes in environmental inputs. Thus, the susceptible child is not irrevocably damaged, but rather is represented as responsive to alternative biomedical technologies (194–96).

The Internet offers biomedical models of becoming that interpellate parents as potential saviors of afflicted children. Internet-based health advice thus empowers everyday people with knowledge, allowing them to assume responsibility for their own care, but simultaneously shifts responsibility for risk away from traditional medical-administrative apparatuses to parents, particularly mothers. Middle-class mothers are especially held responsible for knowing everything there is to be known about their child's condition. The wide availability of knowledge therefore has the effect of making the ill-informed mother appear indifferent to her child's well-being. The openness of the child's condition demands the responsible mother's total involvement in her child's care.

The electronically circulating emergent ontology of the ADHD child as impaired but capable of self-actualizing has evolved into yet another discourse that approaches ADHD as "difference" but not defect or disease (Parker-Pope "A New Face"). A new form of biosociality, or shared sociality based on biology (Rabinow 181–93), has emerged around ADHD, spurred in part by the gold-medal-winning swimmer Michael Phelps, whose mother has spoken publicly about his diagnosis. Ironically, although Phelps discontinued medication at age ten, he now serves as a national spokesperson for McNeil Pediatrics (a pharmaceutical company), which manufactures Concerta (generic methylphenidate). No doubt McNeil Pediatrics sees gain in popular efforts to destigmatize the disorder.

The emergent and increasingly destigmatizing ontologies of ADHD circulating electronically in the early twenty-first century are in some ways incongruent with the rather fixed ontology of the child invoked in the hegemonic, expert-scientific discourse about ADHD and other psychiatric disorders (e.g., autism) codified in authoritative medical texts, such as the *DSM-IV,* and in scientific/medical academic journals. Yet it is the authoritative and scientific-medical framework that is most frequently reported in the popular print media. Newspapers and traditional magazines emphasize studies addressing the genetic basis of ADHD. Genetic accounts do not necessarily discount environmental mediations, but the typical reporting techniques marginalize or elide contributing environmental factors. For example, a 1992 *Scientific American* article reports that the risk for developing ADHD for a child whose identical twin has it is between eleven and eighteen times greater than for a nontwin sibling of a child with ADHD (cited in Barkley). In 2008 the *New York Times* featured an apparent expert on ADHD who stated conclusively that the "heritability of A.D.H.D. is 76 percent" (Baruchin).[3] The strong stress on heritability estimates in mainstream print media accounts implies a kind of inborn immutability that is rarely found in the therapeutic online discourse.

Geneticization of ADHD

As the geneticization of ADHD began to circulate more widely in scientific journals and the mainstream press in the 1990s, behavioral geneticists began to link the disorder to a wide array of "risk" behaviors and risky outcomes. Consider the 2007 headline in the popularly oriented online digest, *ScienceDaily*: "Hyperactivity and Academic Achievement Could Be Linked by Genetics":

> Children who are hyperactive tend to do worse academically than their peers who are not hyperactive. Although the relationship between such behaviors as overactivity, impulsivity, and inattentiveness in children and poor achievement

in math, reading, language, and other areas has been well documented, little is known about the reasons for this link. New research shows that the tie may be due to genetic influences.

The *ScienceDaily* article addresses research by Kimberly Saudino and Robert Plomin (972–86). Arguing that "recent research suggests that genetic factors explain more than 50 percent of the variability in academic achievement," Saudino and Plomin hope to support the hypothesis that "hyperactivity and achievement are associated because of common genetic influences" (973). They acknowledge that social environments "convey risk for developing hyperactive problem behaviors" that may also impact academic achievement, but believe that twin studies allow researchers to accurately measure the relative contributions of environment and genetics to educational outcomes.

Saudino and Plomin contribute to a genetic discourse that constructs genes and environment in a binary formulation that represents the environment purely in terms of social inputs while genes are represented as relatively fixed biological predicates. This discourse prevails in behavioral genetic research and in political discourse that draws upon genetic accounts to justify social policy. Critics challenging this discourse contend it ignores or marginalizes the synergy across genes and environment and ignores or statistically trivializes the potential for exogenous bioenvironmental factors such as lead and nutrition to impact genetic and epigenetic processes during pregnancy and after a child's birth (see Partridge for discussion 985–86). Critics' efforts to contest what they say as the artificial dichotomy between genes and environment find expression in journals such as *Environmental Health Perspectives* but typically fail to be publicized in the popular print media. Vested commercial interests may explain the perseverance of the most simplistic formulations of the relationships among genes, brains, and behaviors (see chapter 3 for discussion).

Behavioral geneticists, biological psychiatrists, and neuroscientists often adopt a kind of biological positivism that sees genes as causing brain states, which in turn cause affect and behavior (see Nadesan *Governing* 162–78). Researchers in these areas hope to map relationships across genes and neural chemistry in order to explain individual differences in impulsivity, aggressiveness, intelligence, and so on. Accordingly, the mainstream expert discourse of ADHD in the early twenty-first century tends to view inborn genetic influences as causing abnormal brain chemistry. Abnormal brain chemistry is typically explained in relation to disruptions in the production or uptake of dopamine, a neural transmitter. The preferred technology for manipulating dopamine levels within this explanatory frame is, of course, pharmaceutical. Methylphenidate still remains the drug of choice, although twenty-first-century bioscience promises new gene-targeting medications.

Medication is the most commonly employed strategy for governing brain states. Some mainstream medical researchers therefore view a failure to medicate as a public health problem. For instance, in 2006, *ScienceDaily* reported that one study found that only half of children with diagnoses of ADHD used medication. The researchers cited in the article express concern about lack of medication:

> "From a clinical point of view, this study affirms that for whatever reason, many children who could benefit from treatment are not receiving it," says first author Wendy Reich, Ph.D., research professor of psychiatry in the William Greenleaf Eliot Division of Child Psychiatry.
>
> It is possible those children aren't being identified at schools or pediatricians' offices or that their parents are choosing not to put their children on stimulant medication, according to Reich.
>
> "It may be that mental health professionals need to do a better job of explaining the risks and benefits of treatment," Todd says. "The vast majority of parents whose children were involved in this study reported that their kids improved with medication, and when used properly these drugs have been shown to be very safe." ("Almost Half")

The article's title, "Almost Half of Kids with ADHD Are Not Being Treated," implicates methylphenidate as the *most important* treatment for ADHD, despite recent research calling into question the drug's long-term effectiveness and its rare but significant "side effects," including stunted growth, heart difficulties, and sudden death from stroke (Favole D6; Vendantam "Debate Over"). Perhaps these rarely mentioned side effects explain why atypical antipsychotics such as Risperdal, Zyprexa, and Seroquel are now being used to "treat" children diagnosed with ADHD and those with similar symptoms grouped under the newly vogue classification of bipolar syndrome (Carey "Bipolar Illness"; Tomsho D1, D6). However, these drugs also present significant side effects, including massive weight gain and zombification.

The idea that behavioral deviations are a direct result of chemical brain imbalances seduces the twenty-first-century biomedical imagination (see D. Healy *Creation* 1–391). ADHD, autism, depression, obsessive compulsive disorder, anxiety disorders, conduct disorders, and bipolar disorder are similarly described in terms of chemical brain imbalances. Hypothetically, the range of environmental inputs potentially impacting brain chemistry seems unlimited. Diet, social stress, physical activity, and sleep levels all seem to be good environmental candidates capable of shaping production levels and uptake of the brain's neurochemicals. Yet these environmental contributions are minimized or excluded from the mainstream expert discourse of brain chemistry's impact on behavior and emotional states. Most importantly for this discussion, the role of undue lead exposure in producing ADHD/ADD-like symptoms has almost disappeared from mainstream accounts despite growing epidemiological links between lead exposure and ADHD diagnoses ("Attention Deficit May Be" D3; Grandjean and Landrigan 2167–78; Levin et al. 1285–93; Needleman 47–54).

At the turn of the twenty-first century, as genes are cast as disproportionately shaping neural chemistry, what is stressed is the inborn genetic contribution to ADHD and similar disorders. While genes do in fact regulate brain chemistry, the hegemonic model of genetic operations is mechanical and thereby discounts how genetic expression is mediated by environmental inputs and synergistic processes. The next section explores the emergence and growing dominance of the dogma of the gene.

The Dogma of the Gene: Genetic Susceptibility, Biocapital, and the Biogenetic Governance of Childhood

Genes are merely stretches of DNA. Humans share nearly the same DNA sequence: 99.9 percent of one individual's DNA sequences will be identical to those of another person. Of the 0.1 percent difference, over 80 percent will be single-base substitutions (SNPs). Genes shape every aspect of corporeal existence, but genetic operations are rarely mechanically determined. Rather, they are shaped by chemical cues that impact how genes produce proteins. Insel and Collins observe that

> by alternative arrangements of RNA following transcription of the DNA, 30,000 genes can code for 100,000 proteins. Adding posttranslational modifications (i.e., changes to the protein following translation from RNA) like proteolysis, phosporylation, and glycosylation may ultimately yield as many as 1,000,000 different human proteins. (617)

The narrative of genetic dynamism, flexibility, and openness is being written by twenty-first-century genomic science but has yet to circulate widely within the popular media, nor has it impacted the dogma of behavioral genetics. Behavioral genetics and popular print media accounts are particularly locked in the Mendelian universe of genetic causality.

The philosophy of genetic determinism was promulgated in early press accounts introducing the public to the new science of the gene. A 1958 article published in the *New York Times* titled "Small Wonder Called the Gene" is significant because it helps explain how popular understanding and dialogue about genetics were framed from the early 1950s onward. Accordingly, the article begins by introducing readers to the idea of the gene:

> Tiny bits of matter, barely visible under powerful microscopes but present in every human being, suddenly have assumed monumental importance in world affairs. These pieces of protoplasm contain genes, the molecules which carry the traits of mankind—our virtues and our faults—from generation to generation. (Springer 15)

The article then goes on to explain the significance of genes:

> Genes are responsible for more of our physical, mental and emotional makeup than most of us realize. For example, the union of genes at the moment of conception of any one of us determined the color of our eyes and skin, the size and shape of our facial features, the length of our limbs and fingers, the general level of our mental capacity. (Springer 15)

Genes, the author instructs, predict personal attributes:

> Although there is a breathtaking element of chance—and always the possi-
> bility of another Leonardo da Vinci or William Shakespeare in every human
> conceived—students of eugenics can often study the backgrounds of parents
> and predict with fair success the general intelligence of their children. Many
> psychologists have conducted tests to prove that children with brainy parents
> have higher I.Q.'s, on the average, than youngsters of fathers and mothers who
> are but normally intelligent. (Springer 24)

The dangers and benefits of genetic heritability are illustrated by contrasting
the (presumed) hereditary musical genius of the Bach family in Europe with
the "infamous Jukes, a New York family with a fantastic record of criminality, disease and imbecility." The article states that a sociologist, Richard
L.Dugdale, obtained precise information on 709 descendants of two Juke
sisters and found that 140 had been imprisoned and 280 and had been paupers. In seventy-five years, he estimated, the state spent $1,308,000 to care
for them (24–25).

This account represents genes as fixed and determining entities that play
the most significant role in predicting individual differences measured in
psychological inventories. It draws upon highly charged examples to demonstrate its arguments. This article's logic continues to prevail as the hegemonic frame shaping mainstream print media's accounts of genetic research,
as illustrated by the following representative articles:

- "Back to Genes: More Evidence to Suggest that Genetic Endowment
  Limits, Without Destroying the Possibilities of Social Change" (11)
- "The Role of Genetics in IQ Scores" (Rensberger and Hilts 9)
- "Scientists Link Anxiety to Specific Gene" (Talan A8)
- "Researchers Find Stress, Depression Have Genetic Link" (Vedantam
  A10)
- "Possible Link of Violence, Gene Found" (Cooke A18)
- "Bullying Behavior, Blame it on Bad Genes" ("Bullying")

For at least three decades, popular press accounts repeated the dogma that
genes explain personality traits and attributes: "Whatever it is that IQ (intelligence quotient) tests measure, it appears to be something that is to some
extent genetically transmitted, according to a study described as the most
comprehensive yet on the subject" (Rensberger and Hilts 9). The popular
print media have been particularly intrigued by the heritability of undesirable or risky traits or behaviors, as illustrated by a 2006 article in the
New York Times titled "That Wild Streak? Maybe it Runs in the Family"
(Harmon). Although the article begins with risky behavior such as fast driving and skiing, it eventually extends its discussion of heritability to include
violent behavior and raises questions about how risky gene holders ought be
treated: "Already, some scientists suspect a specific gene plays a role in violent
behavior, for instance, and a discussion has already begun over how people

bearing such genes should be treated" (Harmon). The U.S. Department of Health's 1990s-era "Violence Initiative," which applied organic psychiatry and behavioral genetics to the problems of violence and criminality (Allen 1–24), helped stimulate this type of research on heredity and criminality.[4]

Why are simplistic genetic explanations so pervasive in popular accounts? On the one hand, the narrative of mechanistic genetic determinism is simple, and the public has been conditioned towards its receptivity. On the other hand, the model of mechanistic genetic action offers opportunities for capitalization that have shaped genetic problem-solution frames and research funding in the early twenty-first century.

The most obvious commercial gains linked to genetic accounts involve pharmaceutical interests. Pharmaceutical and biotechnology companies have a vested interest in genetic accounts of "mental illness" because these accounts promote biological problem-solution frames. Molecular/genetic understandings of mental illness invite molecular/genetic interventions (see Healy *Creation* 1–390). Psychiatric drugs are big business, and sales of psychiatric drugs are very profitable. The invention of any new "biological" psychiatric condition is likely to stimulate pharmaceutical drug sales. For instance, the explosion of bipolar diagnosis accompanied a 12 percent surge in sales of antipsychotic drugs between 2006 and 2007 (Tomsho D1).

New markets for pharmaceutical drugs involve medicalizing and destigmatizing psychological conditions such as stress, anxiety, and depression. New markets can also be created by narrowing the parameters of psychic and behavioral normality through expert psychological and psychiatric discourses (see Nadesan *Constructing* 132–35). All of these techniques— medicalization, destigmatization, narrowing—have contributed to the production of amazing prevalence statistics for mental illness, as illustrated by a recent study that reportedly found that one in five young adults suffered from "a severe personality disorder" ("Study: 1 in 5 Young Adults" A4). The 2003 President's New Freedom Commission on Mental Health Report claims that between 5 to 9 percent of U.S. children suffer from "a serious emotional disturbance" (New Freedom Commission; see T. Miller 75–104). The report advocates frequent, regular screenings of adults and children for signs of mental illness, followed by prompt, evidence-based treatment.

A novel approach to detecting childhood differences requiring normalization has recently emerged. This novel approach, funded by the federal government, aims to create standards of neurological normality. Researchers are using MRI brain scans to map the profiles of 400 children. Norms will be developed from the MRI data:

> About 400 healthy newborns to teenagers, recruited from healthy families, are having periodic MRI scans of their brains as they grow up. They also get a battery of age-linked tests of such abilities as IQ language skills and memory. The project is funded by the National Institutes of Health. The MRI images measure how different parts of the brain grow and reorganize throughout childhood. Overlap them with the children's shifting behavioral and intellectual abilities at each age, and scientists expect to produce a long-sought map

of normal brain development in children representative of the diverse U.S. population. ("Researchers" A19)

Neurological normality provides a novel grid of intelligibility for identifying, mapping, and measuring differences requiring normalization.

Although destigmatization of mental illness can prompt individuals to seek help for themselves or their children, it also encourages a biopolitics of surveillance and risk management that can medicalize both normal differences within the population and environmental-produced psychic anxiety. Moreover, lack of adequate private and public funding for integrated mental health treatment protocols often produces overreliance on pharmaceutical management of troubling "symptoms," leading to the potential zombification of children on antipsychotics, among other troubling effects (see T.Miller 75–104). Medicalizing human differences and environmentally produced disturbances fosters pharmaceutical profits while potentially posing risks to children diagnosed as different.

There are other revenue streams that derive from medicalized conditions beyond those associated with pharmaceuticals. In recent years, private and university research laboratories have produced new revenue streams by patenting genetic materials purportedly linked to biological disease conditions. Indeed, early twenty-first-century biomedical research on ADHD has arguably been at least partially driven by patent opportunities. Researchers look for genes implicated in the disorder that can be patented and they also seek to patent procedures for studying ADHD neurochemical processes. In 2006 a dopamine transporter gene was purportedly linked to the heritable transmission of ADHD (Ohadi et al. 1–4). Also in 2006, the U.S. Patent Office granted Madras et al. Patent 7,081,238 for their methods for "diagnosing and monitoring treatment of ADHD by assessing the dopamine transporter level" (http://patft.uspto.gov/netacgi/nph-). The method presumes "an elevated level of dopamine transporter in the patient is indicative of ADHD." Likewise, the U.S. Patent Office granted Bunzow et al. Patent 6,783,973 for "Mammalian Catecholamine receptor genes and uses." The researchers claim these receptors may be a contributing factor in Parkinson, schizophrenia, and ADHD (http://patft.uspto.gov/netacgi/nph-). Patent 6,660,476 was issued to Comings et al. in December 1993 for "Polymorphisms in the PNMT gene" such that "the method of claim 1, wherein said disorder is attention deficit hyperactivity disorder and said predetermined risk for said disorder is an increase in double heterozygosity at the nucleotide positions -387 and -182 of the PNMT gene" as determined by the researchers (http://patft.uspto.gov/netacgi/nph-).

As Nikolas Rose observes, "Medical truth itself, has become subject to intense capitalization" (*Politics* 41). Genes have become transformed into commodities, shaping research trajectories and problem-solution frames (Nelkin "Molecular" 558). The idea of genes as commodities encourages determinism, because causative genes have greater commercial value. Both researchers who wish to study patented genes and pharmaceutical companies wishing to manipulate their operations must pay patent holders. Margaret Lock expresses concern over the assumptions implicit in this drive to patent genes implicated in medical susceptibility: "We now know, DNA segments are rarely, and possibly never, *determinants* of disease," yet the language of genetic causality prevails and shapes understanding of self and others (64). Risk for a psychiatric condition or disease is given definitive calculations in genetic discourses, while the role of other risk factors is rendered background static.

In an essay titled "Behavioral Genetics and Dismantling the Welfare State," Dorothy Nelkin illustrates how the discourse of genetic determinism is employed in policy and legal debates to support claims about the limits of social intervention (157). Nelkin describes how antisocial behavior has been geneticized in popular media accounts, encouraging public belief in the existence of "bad" genes (158–59). Individuals who inherit bad genes may be viewed as having less volitional control over their antisocial behavior, perhaps mitigating courtroom sentencing for crimes, yet are viewed as less responsive to social engineering. This logic was evident in media reports of a mutation of the *MAOA* gene that was initially linked with violent behavior. Subsequent research argued that those who have the mutation are only prone to violence when raised in environments characterized by aggression, violence, and deficits of parental care (Begley "Life Events" B1; Kim-Cohen et al. 903–13), whereas other studies called into question any relationship between the gene and aggression (Morris et al.; Ossorio and Duster 115–28). However, subsequent studies have typically received less media attention and often fail to dispel false claims. Consequently, the public is led to believe that individuals are genetically predestined towards particular conditions or behaviors. Educational failure and success are both cast as genetic, as are alcoholism and poverty. Social interventions can be viewed as doomed to failure or as having limited impact. Behavioral geneticists Robert Plomin and Ian Craig imply this conclusion when contending that "individuals select or create environments that foster their genetic propensities" (s42). These ideas contrast with the belief, in the 1930s, that character malleability could temper the heritable deficiencies of children.

The geneticization of personality, behavior, and cognition has significant implications for the government of childhood. In particular, it has spurred efforts to find and patent techniques for genetically screening and sorting embryos and fetuses. This biopolitical sorting of fetuses has recently expanded to include efforts to sort children according to (believed) inborn talents, abilities, and liabilities. The biopolitical sorting of fetuses began with prenatal testing for Down syndrome but now includes much more sophisticated means for detecting genetic defects capable of producing "risks" for disabilities. A new method called comparative genomic hybridization was patented in the 1990s to detect sequence deletions, duplications, and translocations in targeted areas of an individual's genome (U.S. Patents 5,665,549 in September 1997 and 5,965,362 in October 1992). The abstract for Patent 5,965,362 explains:

> Disclosed are new methods comprising the use of in situ hybridization to detect abnormal nucleic acid sequence copy numbers in one or more genomes wherein repetitive sequences that bind to multiple loci in a reference chromosome spread are either substantially removed and/or their hybridization signals suppressed. The invention termed Comparative Genomic Hybridization (CGH) provides for methods of determining the relative number of copies of nucleic acid sequences in one or more subject genomes or portions thereof (for example, a tumor cell) as a function of the location of those sequences in a reference genome (for example, a normal human genome) . . , Amplifications, duplications and/or deletions in the subject genome(s) can be detected. (http://patft.uspto.gov/netacgi/nph-)

The patent description explains the method's significance: "An important utility of CGH is to find regions in normal genomes which when altered in sequence copy number contribute to disease, as for example, cancer or birth defects."

In 2007 the World Intellectual Property Organization granted PCT/US2007/071247, Diagnosis of Fetal Abnormalities Using Polymorphisms including Short Tandem Repeats (http://www.wipo.int/pctdb/en/wo.jsp?WO=2007147073). CGH is currently being used for prenatal diagnostics to identify fetuses potentially susceptible to particular syndromes caused by, or linked to, deletions or additions of genetic material (Stein "Fetal Gene" A8). Concern already exists that this type of genetic screening could be used to identify susceptibilities for a wide range of disorders linked, but not caused by, genetic factors. CGH transforms a child's risk into suspect absences or duplications in genes of alleles prior to his or her birth. Risk becomes the punctuation of one's genome; missing or additional sequence elements produce risky or dangerous individuals.

Parents, fearing these risks, may make reproductive decisions based upon them. Parents are responsibilized for producing perfect children, as described so poignantly by Joan Rothschild in *The Dream of the Perfect Child*. Risky genes raise the apparition of disability, potentially terrifying expectant parents. Given the lack of social welfare supports for parents and the necessity of two-income households, the decision to select against risk is understandable. Still, public reaction to and consumption of prenatal genetic screening devices in the United States are mixed. Social conservatives largely reject these technologies as violating God's design. Disability rights activists have voiced tremendous concern about prenatal testing and have organized using the Internet to disseminate their concerns. Environmentalists have also raised an alarm, citing the role of environmental pathogens in shaping genetic expression (i.e., epigenetics).

Despite oppositional discourses and movements, the desire to optimize children's outcomes has encouraged some parents to embrace the gospel of the gene. Indeed, some parents believe that knowledge about their children's genetic makeup can improve parenting. Each child is believed to be unique within this framework. Because of this uniqueness, no single set of experiences or parenting techniques works for every child (Schwarz 883). Genetic testing can help parents provide the "right" environmental conditions for their child's unique genetic constitution. Parents also hope that genetic testing might alert them to a child's inborn hypersusceptibility to chemicals and pollutants, enabling them to take individualized action. Genetics might even predict talents such as music or sports. Capitalizing on this belief, a new genetic test has been developed that screens children for athletic ability, as the *New York Times* reports:

> In health-conscious, sports-oriented Boulder, Atlas Sports Genetics is playing into the obsessions of parents by offering a $149 test that aims to predict a child's natural athletic strengths. The process is simple. Swab inside the child's cheek and along the gums to collect DNA and return it to a lab for analysis of ACTN3, one gene among more than 20,000 in the human genome. The test's goal is to determine whether a person would be best at speed and power sports like sprinting or football, or endurance sports like distance running, or a combination of the two. (Macur)

A parent interviewed for the story opines that the test will "relieve a lot of parental frustration." Parents wealthy enough to afford the test have yet another technology at their disposal promising to help optimize their children's successes. Genetic tests' predictive utility matters less than the illusion of control they bestow.

Those critics who view the environment as mediating biology reject this type of genetic fortune-telling. Environmental critics charge that genetic testing and the discourse of genetic proclivity/susceptibility gloss the real uncertainty surrounding the relationship between most diseases/traits and genes, while marginalizing the role of social and biochemical environmental mediations. Moreover, critics worry that genetic accounts will further erode efforts to identify environmental "risks" to human health and well-being. Although the discourse of genetic susceptibility could embrace environmental mediations, a variety of factors have converged to privilege agentive genes over mediating environments. Robert Proctor explains how genes have trumped environment in a new kind of biological determinism:

> One of the dangers of biological determinism, then, is that the root cause for the onset of disease is shifted from the environment (toxic exposure) to the individual (genetic defects). The scientific search shifts from a search for mutagens in the environment to biological defects in the individual... *Consumer Reports*, in its July 1990 cover story on genetic screening, warned, "The danger is that industry may try to screen out the most vulnerable rather than clean up an environment that places all workers at increased risk." (Proctor 80)

The danger is that emphasis is "placed on defects in the individual rather than defects in the industrial product or environment" by locating the susceptible individual front and center (80). This framing can have shape social policy by encouraging individualized technologies of the self for reducing risk. Personalized risk management approaches are consistent with neoliberal logics of government.

Neoliberal approaches to risk assessment complicate efforts to regulate potential environmental hazards. For instance, Ronald Reagan, who administered the United States' first, fully neoliberal presidential administration, required OSHA (the Occupational Safety and Health Administration) to relax safety requirements concerning worker exposure to lead (Medley 71). Regulatory approaches in the United States presume chemicals to be safe unless proven otherwise. New regulations are difficult to enact and require conclusive evidence that chemicals threaten human health and that the threats posed to human health exceed the costs of new regulatory controls. For example, decades of research on deformities in aquatic life were required before the EPA finally ruled in 2009 that pesticide manufacturers will be required to test (only) sixty-seven chemicals to see whether they disrupt the endocrine system (Eilperin A1). It will take years of laboratory testing before risk assessments are finalized. Regulatory action will take even more time and will occur only with clear evidence of harms. The discourse of genetic susceptibility to environment harms complicates efforts to establish that the costs of health effects outweigh regulatory costs because sensitivities to environmental chemicals differ across individuals and are mediated by a seemingly infinite range of variables, including age of exposure and nutrition. Consequently, neoliberal calculi for assessing and regulating risk are likely to privilege the interests of polluters and chemical producers given the difficulties of establishing significant effects across the population and the relative ease with which producers/polluters can demonstrate the costs of enhanced regulations. Actuarial risk assessment of genetic susceptibility may therefore displace responsibility for reducing exposure to privatized individuals.

Affluent populations within the United States have greater awareness of environmental "risks" and have the resources to adopt personalized strategies for reducing their exposure to environmentally risky substances. They live in suburbs distanced from industrial polluters. Their children are less likely to attend the 435 schools known to be directly exposed to industrial pollutants dangerous to students' health (Morrison and Heath). Affluent parents can breastfeed or purchase healthier consumer products when they read in the press that perchlorate from rocket fuel laces powdered infant formula (Sapien and ProPublica) and that more than half of baby toiletry products are laced with carcinogenic substances (Layton A4). Not surprisingly, organic produce and milk have grown in popularity among affluent consumers. Organic laundry services and pest-control services have also emerged as ecofriendly alternatives. A 2008 article in the *New York Times*, "For 'EcoMoms,' Saving Earth Begins at Home," suggests that a new kind of "ecoanxiety" permeates middle-class culture (Brown). Although the article links ecoanxiety to a desire to "save earth," it is likely that a more potent motivator is fear for one's offspring. Aspiring, affluent parents adopt many "technologies of the self" aimed at minimizing environmental risks to their children. This capacity—born of affluence and education—to adopt personalized risk management strategies may undermine support for widespread government regulation.

However, affluence and education alone have proven insufficient protections as more research studies implicate ubiquitous and unavoidable chemical risks. Parents learned that plastics used in water bottles and the lining of canned products contain a potential hazard called bisphenol A (Tugend). News stories about phthalates cause panic among environmentallyactivated parents since these ubiquitous chemicals, which are often found in the plastics used in infant bottles and children's toys, may be linked to developmental problems in children because of their potential role as endocrine disruptors (Pereira B1). Growing concern about widespread and unavoidable health risks to children may eventually erode confidence in the market's capacities for self-regulation. The U.S. experience of allowing corporations to self-police (i.e., self-regulate) the safety of their products has resulted in the proliferation of chemicals of questionable safety for human development. Many environmentally aware parents now desire a tighter regulatory regime. However, as will be explained in the concluding chapter of this book, the regulatory response to the financial crisis that began in late 2007 has not proved promising in this regard.

In sum, the increasing popularity of genetic accounts of human behavior, abilities, and disease, coupled with the individualization of risk management, undermines public demands for social welfare supports and regulatory action. The use of genetic explanations to explain differences in children's academic outcomes and aggressive tendencies are contributing to a .growing sense that social welfare interventions for at-risk children have limited benefits, given inborn predispositions.

CONCLUSIONS

Turn-of-the-twentieth-century research on children's individual differences is driven by commercial considerations that encourage genetic determinism. Unlike genetic explanations, environmental explanations of measured differences in aptitudes or intelligence offer few opportunities for commercial exploitation. Environmental regulations cost industries and therefore face strong opposition. Neoliberal risk assessment procedures favor quantifiable measures of risks and costs and therefore devalue the contingent/synergistic calculations of environmental risk calculations. Environmental risk-reduction policies and programs are favored only when quantifiable benefits are perceived as outweighing financial costs or when the risk-reduction policies can be exercised by individuals through personalized technologies of the self.

Poor children living in older housing and near industrial production continue to suffer from undue lead exposure, but the popular imagination has largely forgotten their plight, beguiled as it is by the threat of contaminated toys from China. While contaminating toys are not to be disregarded for the perils they might pose, they simply do not have the same epidemiological impact as leaded paint dust. Yet environmental harms to poor children do not seem to figure significantly in societal cost-benefit risk assessments, undermining environmental and social activists' efforts to step up environmental protections. Furthermore, crowding out images of the vulnerable child at risk for lead poisoning is a new image of the lead-burdened, "risky" child who threatens public safety.

Today, environmental accounts of cognitive and health outcomes must compete with biogenetic formulations promoted and backed by many of the nation's scientific authorities. ADHD and a variety of "heritable" learning disorders now explain school failure and delinquency in the pages of academic journals and science magazines. Consequently, the children who bear the greatest statistically defined "risk" for lead exposure are increasingly cast as ontologically risky by virtue of their bad genes, heritable susceptibility, and statistically defined cognitive limitations. This formulation encourages and legitimizes declining commitments to redistributive and protective policies toward children's environmental and social living conditions.

The aspiring middle and upper classes have seized upon genetic accounts but are likely to stress their own children's susceptibility rather than genetic fate. Many affluent parents have become hypervigilant about environmental perils that potentially threaten their unborn and born children. Adopting consumer-driven technologies of the self, these parents are likely to purchase organic foods and chemically safe products to optimize their offspring's success. Diagnosis of a learning disorder or disability is likely to bolster parental efforts to compensate for their children's disability through the consumption of specialized learning products, food and vitamin regimes, and tutoring. Affluent parents demand and receive specialized services for their children in public schools fearing legal action. These activist technologies of optimization do not necessarily translate into social-environmental support for all children.

Although growing concern about widespread and unavoidable environmental harms—ranging from toxins to global warming—could encourage publics to embrace regulatory logics, the handling of and responses to the economic crisis that began in 2007 suggest that industry needs may very well be privileged over all other considerations. The last chapter of this book will explore these developments further. Discussion now turns to examine a new security discourse that has emerged to classify and govern poor children globally.

CHAPTER 5

Biopower, Security, and Development

Biopolitics approaches population as a political and scientific problem-space. Contemporary biopolitics operates primarily, although not exclusively, through security mechanisms rather than disciplinary ones. This chapter specifically addresses how populations, particularly childhood populations, are constituted in political and social policy as national and international security concerns. The chapter explores two specific childhood populations. First examined is the growing population of poor children living in a politically aware, but disenfranchised, developing world. Examined next are the children of migrant flows into the United States over the last several decades and how cultural imaginings and governmental technologies have approached this population. These discussions are contextualized in relation to the idea of the United States as a "security state."

Analysis of the United States as a "security state" may at first appear incongruent with the types of Foucauldian analyses pursued in this project. Foucault preferred to analyze heterogeneous technologies of government that discipline and securitize populations over the centered power of sovereign authority. In *Security, Territory and Population*, Foucault links sovereignty with mercantile control over a specifically delimited territory in contrast to the disciplinary normalization of industrializing societies and the governmental biopolitics of contemporary societies (11). He argues in *Society Must Be Defended* that while the ancient right of sovereignty was defined in relation to "the right to take life or let live," in the modern period sovereignty became "the right to make live and to let die" (241). The right to make live and let die is not disciplinary but rather entails a " 'biopolitics' of the human race" (243). Biopower operates through security technologies such as vaccination, aimed at regularizing and optimizing life (246–49). Although biopower circulates, many of its strategies of deployment originated with the state working in concert with a whole series of "sub-state institutions" such as medical institutions, welfare apparatuses, insurance, and so on (250). Working in these substate and private biopolitical institutions, contemporary petty sovereigns rely upon force and execute violence in the course of everyday life (see Chappell 313–34). The power of petty sovereigns is unmistakable, but this focus on the micropolitics of power in institutions and across everyday life should not preclude analysis of state power. In the modern era, state power is often cloaked or rationalized by biopolitical objectives.

According to Foucault, the state operates in a "biopower mode" when it seeks to cultivate and securitize the life forces of its populace; this process entails eliminating biological threats to its populace (*Society* 256). War and conflict between states operating in the biopower mode rely on and promote racism in the sense that enemies are posited as posing biological threats to the state's home population or "race" (257). Racism in the modern sense, Foucault, argues is bound up "with the workings of a State that is obliged to use race, the elimination of races and the purification of race, to exercise its sovereign power" (258). Foucault uses the Nazis to illustrate how state sovereignty is exercised in the biopower mode; however, his analysis of this modality of power can apply to any state that defines security biopolitically and then acts violently in its defense. In essence, state sovereignty persists in the modern period, although power is rationalized in relation to efforts to securitize populations rather than in relation to territorially linked divine rights. To repeat, state power is rationalized in relation to internal and external "security" threats.

This approach to understanding state power biopolitically is taken up by David Campbell in *Writing Security*. Campbell argues that "the project of securing the grounds for identity in the state involved an 'evangelicalism of fear' that emphasized the unfinished and endangered nature of the world" (61). This passage illustrates Campbell's argument that states' efforts to foster national identities usually involve articulations of external enemies seen as endangering established or preferred ways of living. This process of positing identity oppositionally can also entail nonstate others who are perceived as posing some threat to the life or vitality of the state's populace. The creation of a national identity mobilized around commonly perceived security threats helps legitimize state power and the execution of state repressive forces.

Past and present formulations of the identity of the United States have relied on articulations of common foreign and domestic enemies, ranging from suspect foreign powers such as Japan and the former Soviet Union to domestic "threats" such as Native Americans, Communist infiltrators, and so on. Domestic enemies have been understood as polluting the nation through their deficiencies or through the contaminating influences of their lifestyles, while foreign enemies threaten national vitality by consuming scarce resources or by threatening the "American way of life" (see Nadesan *Governmentality* 183–210). Efforts to neutralize the "threats" posed by domestic enemies have included disciplinary enclosures, targeted surveillance, and normalization. Efforts to neutralize foreign "threats" have included the direct means of war as well as the more indirect and pastoral interventions of foreign aid and philanthropy. In other words, U.S. state apparatuses have drawn upon a variety of modalities of power, including circulating security apparatuses (i.e., government), discipline, and sovereign power, in the process of producing and responding to "enemies" or threats to the welfare of the state. These deployments are consistent with Foucault's claim in the essay "Governmentality" that sovereignty, discipline and government do not in fact constitute a series of historical displacements (102).

This chapter will first introduce a historicized look at the characterizing features of the U.S. security state in the twentieth century. This formulation of the U.S. security state will then contextualize discussion of early twenty-first-century efforts by the United States to securitize its populace and "way of life" from the biopolitical threats purportedly posed by foreign childhood populations. Representations of foreign children as both dependent and threatening point to the fusion of biopolitics and state sovereignty framing this chapter's analysis.

The U.S. Security State

The Cold War played an important role in shaping the U.S. evolution toward a security state. On March 2, 2006, Clifford A. Kiracofe Jr., a former member of the U.S. Senate Foreign Relations Committee, gave a speech in Berlin titled "U.S. Imperialism: The National Security State." Kiracofe begins his speech by describing what he perceives to be the characteristic features of the U.S. security state:

> The project for the imperial Presidency, garrison state, and imperial foreign policy, was advanced after World War II by Presidents Harry Truman, Lyndon Johnson, and Richard Nixon. For five decades, the project has relied on the manipulation of fear, and the creation of "emergency" conditions, through the systematic deception of the United States public and Congress about the international situation and foreign threats.

These features—"the imperial Presidency, the garrison state, and imperial foreign policy"—have arguably shaped everyday life and the government of American children and childhood populations abroad.

Kiracofe attributes the origins of the U.S. evolution toward a garrison state to the Korean War, which triggered escalation of the Cold War.[1] Harold Lasswell coined the idea of a "garrison state" in 1941 to describe a "world in which the specialists on violence are the most powerful group in society" (455). Lasswell observes that creation and maintenance of the garrison state require a "deep and general sense of participation in the total enterprise of the state" by the populace (458). One way this participation is fostered is through the populace's acknowledgment of its absolute vulnerability. Lasswell suggests that aerial warfare, which abolished distinctions between civilian and military functions, enhanced this sense of collective vulnerability in the modern era (459). Nuclear weapons created the potential for entire populations to be eliminated. It is therefore not surprising that the Cold War mentality infused nearly all aspects of everyday life and popular culture.

The unprecedented U.S. military-industrial complex that prevailed from the Cold War (a.k.a. "military Keynesianism") through the contemporary period found justification in the perils of nuclear warfare and the domino effects of "encroaching communism," which threatened annihilation of "democratic capitalism." Garrison logics drove post–World War II expansion of military spending, creation of civil militias, enforced patriotism in mandatory cultural displays (e.g., pledge of "allegiance" at schools), and the centralization of executive power in the form of the "imperial Presidency."[2] The declared end of the Cold War, with the fall of the Berlin Wall and the economic collapse of the former Soviet Union, failed to eliminate the garrison mentality, despite its inconsistency with libertarian neoliberal ideals. Contemporary fears of nuclear and bioterrorism have reinvigorated the garrison mentality, resulting in the proliferation of government and civil efforts to police populations (in the authoritarian sense) domestically and abroad. Chalmers Johnson notes that as of 2001, the Department of Defense acknowledged at least 725 U.S. military bases outside the United States (4). At home, the creation of the Department of Homeland Security—a warlike administrative apparatus that centralizes otherwise diffuse police and military apparatuses—illustrates the post–Cold War growth of domestic garrison institutions. Likewise, the development of bioweapons labs and research centers throughout the nation illustrates the way the garrison logic has captured purported engines of economic innovation such as twenty-first-century bioscience.

Unlike Lasswell, Foucault focuses on modern security apparatuses that operate through more pastoral biopolitics by eliciting and shaping the health of the population. Yet, as explained above, Foucault did address how states operating in the biopower mode deploy their capacities to kill and let die. The U.S. state's actions within the biopower mode illustrate both types of security deployments. As developed in chapter 2, the U.S. Cold War security state incorporated a variety of proto-eugenic strategies aimed at enhancing the biovitality of the population, which operated primarily on women's sexuality and the government of childhood.[3] The U.S. Cold War security state also incorporated a variety of symbolic and practical strategies aimed at eliciting patriotic citizenship, promoting military service by valorizing combative "American" masculinity,[4] and directing military-industrial engines of growth.[5] These deployments eventually encouraged opportunities for formally demonstrating the military strength of U.S. security apparatuses, including the Korean War, the Vietnam War, the Gulf War, and the Iraq and Afghanistan wars. Preservation of the "American way of life" has legitimized sovereign expressions of U.S. repressive (security) apparatuses for the last sixty years, encouraging the U.S. trajectory toward a garrison state despite the contemporaneous extension of neoliberal logics of economic government.

The formalized threats governing the garrison trajectory have shifted across time. The Cold War U.S. security state gradually evolved into the post-9/11 security state, characterized by the xenophobic "clash of civilizations" legitimized in the works of Samuel Huntington ("The Clash" 22–49) and preemptive warfare, articulated by the former vice president Dick Cheney's "One Percent Doctrine," which legitimized preemptive action against foreign nations or peoples if there existed a 1 percent chance that terrorists could attain "weapons of mass destruction" ("America's Longest" 22). The American state today exercises its might while operating in the biopower mode to "protect and secure" the American way of life from heterogeneous foreign enemies whose geographic dispersions and mobilities seem to require U.S. military responses across the globe.

In the American popular imagination, collective fears and anxiety about a clash of civilizations have been promoted by media-promulgated fantasies of terrorists subversives. However, although U.S. enemies are represented symbolically as linked to the forces of darkness, a closer look reveals their geopolitical proximity to scarce resources. Like all "great powers," the United States has a history of imperial foreign policy designed to secure access to natural resources. The literature on U.S. imperialism is vast, but for the purposes of this chapter what is important is growing global resource scarcity, particularly of oil and precious metals. The U.S. "war on terror" fought primarily but not exclusively in Iraq and Afghanistan is fundamentally driven by resource insecurities, particularly related to oil and gas (see Engdahl 1–245; P. Escobar "Pipeline-Istan"). The symbolic discourse of the war on terror invoked in Huntington's "clash of civilizations" obscures the material bases of the war but appeals to a U.S. public conditioned from childhood by Manichaean rhetorics (see chapter 3). However, the war on terror is but one strategic trajectory in U.S. imperial policy. International lending, foreign aid, and philanthropic-sponsored development projects have also played important roles in the promotion of the U.S. security agenda, "merging development and security" (Duffield 1–265).

The world's impoverished populations are represented within this new security discourse as posing migration, population, resource, and environmental "risks" to the world's affluent populations (see Cooper 90–95). As will be explained below, poor populations are seen as draining resources vital to U.S. biosecurity. Additionally, they are increasingly viewed as radicalized antimodernists who reject democracy and liberalism and thereby threaten the security of the global order. Since these impoverished populations cannot be eliminated through war, the world's most powerful nation-states struggle to contain the threats posed by their needs and frustrations through lending and foreign aid.

For the last thirty years, lending and foreign aid by the United States and global governance institutions such as the World Bank and the International Monetary Fund (IMF) have been guided by neoliberal logics requiring developing nations to liberalize their markets and balance their budgets (Harvey 1–223). Economic liberalization, export-oriented production, and domestic austerity have taken a toll on the world's poorest populations, worsening poverty in some regions and exacerbating gaps between the world's wealthy and poor populations (see M. Davis's *Planet of Slums* 1–207). Private-public partnerships, private investment, microfinance, and microenterprise have gained currency among neoliberal authorities and institutions as the most desirable strategies for redressing global wealth imbalances and the increasing impoverishment of the world's poorest populations (A. Escobar 1–249; Jurik 1–225).

Concern about the world's poor in wealthy nations is largely driven by the purported security risks posed by impoverished populations. Poverty is thus fundamentally tied to security, while poverty alleviation is tied to enhanced market participation for poor populations, optimally achieved through market-directed, for-profit investments and partnerships (a.k.a. "philanthrocapitalism").[6] However, despite much media fanfare and accolades in the developed world, these approaches have not yet proven successful in lifting impoverished populations from the depths of economic despair, nor have these approaches stymied the great flows of peoples driven by dire circumstances to migrate away from their homes.

Population flows have prompted xenophobic responses within wealthy nations. In the United States, the ideology of a nation besieged by hostile foreign agents and powers has been bolstered by right-wing accusations that foreign-born immigrants threaten the national economy and the biovitality of the populace. Anti-immigrant sentiment has directly fueled the growth of right-wing militias in the American West (see U.S. Department of Homeland Security). Furthermore, fears about immigrant-populated gangs and juvenile delinquency in urban areas have contributed to the general public's concerns about security. Anxiety about external and internal "security" threats have encouraged development of a vast network of surveillance technologies and practices within the United States.[7] These domestic threats prompt creation of citizen-warriors who secure domestic and internal security by their "service" to government and privatized security apparatuses. This chapter will explore further how immigrants are represented as posing threats to domestic security after first addressing how poor children abroad are constituted as "security risks" requiring government.

CHILDREN AND INTERNATIONAL BIOSECURITY

National security in the twenty-first century is, as put succinctly by Zbigniew Brzezinski, "inexorably tied to the global condition" (25). Brzezinski claims that maintenance of U.S. security requires "systemic efforts to enlarge the zones of global stability, to eliminate some of the most egregious causes of political violence, and promote political systems that place central value on human rights and constitutional procedures" (25). In 2008 the U.S. secretary of defense Robert Gates called upon the audience at the National Defense University to have an "appreciation of limits" of military power (cited in Barnes). Accordingly, many independent think tanks and military strategists in the United States recommend reconceptualizing security in ways that blur old boundaries between military and civil affairs while embracing risk-management strategies that work less through punitive and disciplinary means and more through pastoral, biopolitical strategies and technologies. Populations generally and children specifically arise as critical concerns within new security discourses.

As Brian Nichiporuk explains in *The Security Dynamics of Population*, this new biopolitics of security must at its base analyze the characteristics and dispositions of population demographics and dynamics:

> As American policymakers stand on the threshold of the 21st century, they tend to view weapons proliferation, hypernationalism, ethnic and tribal conflict, political repression, and protectionism as the principal threats to the open, liberal international order they are trying to create. All of these factors are indeed dangerous and worthy of attention, *but the risks posed to U.S. security interests around the world by demographic factors must not be neglected either. The dynamics of population growth, settlement patterns, and movement across borders will have an effect on international security in the upcoming dec-ad*es, and Washington can do much to solidify its geopolitical position in critical regions by anticipating demographic shifts that have security implications and by working with allies, friends, and international organizations to deal effectively with the causes and consequences of these shifts. (xi, my italics)

Population growth, characteristics, and dynamics are explicitly linked to security. But what kind of biopolitics inheres within, and follows from, the types of linkages and inferences drawn from this type of security discourse?

The biopolitical knowledge and technologies deriving from these linkages emphasize the fertility and mobility of poor, youthful populations. Both Nichiporuk's and Brzezinski's analyses emphasize differences in global fertility rates as playing an important role in the future security of developed and developing nations. Young, fertile populations in developing nations face increasing urbanization, poverty, and resource scarcity, leading to mass refugee outflows, outbreaks of ethnic conflict, and the potential for ideological revolutions. Nichiporuk emphasizes how ideological revolutions could arise in the ring of slums surrounding many large "third world cities" (20). Brzezinski points out that greater literacy and the rise of mass media and technology have amplified risks posed by these poor populations because the "world is now politically awakened to the inequality of the human condition" (42). Both authors emphasize the politically awakened youthful populations' (purported) susceptibility to "the emotional appeals of nationalism, social radicalism, and religious fundamentalism" coupled with a kind of resentment toward the world's economically privileged (Brzezinski 43).

Vast global schisms exist regarding popular awareness of the scope of resource inequalities. Poor populations everywhere are alerted to, or reminded of, their own marginalization as they consume televised images of wealth and privilege, which contrast with the polluted and crowded megacities many of them inhabit. Conversely, the popular imagination in the industrialized world seems to be just becoming aware of the depths and scope of global inequalities. For instance, in the middle of the financial meltdown of 2007–2008, the *New York Times* announced:

> There is a lot more poverty in the world than previously thought. The World Bank reported in August that in 2005, there were 1.4 billion people living below the poverty line—that is, living on less than $1.25 a day. That is more than a quarter of the developing world's population and 430 million more people living in extreme poverty than previously estimated. The World Bank warned that the number is unlikely to drop below one billion before 2015. ("World Bank Finds" A9)

Even Asian nations, those formerly regarded as Asian Tigers, belie their apparent material successes when the Western media inquire further into the plight of their populations, as illustrated by this press report: "The growing gap between rich and poor in booming Asian economies has left behind 'vast numbers of mothers and children,' putting millions of lives at risk, according to a report by UNICEF, the United Nations children's agency" (Ramesh). Climate change and expected global food shortages will exacerbate poverty and migration (see Bourne 26–59; Vedantam "Climate Fears" A1).

Samuel Huntington, among many security analysts employed by Western powers, sees global disparities as exacerbating the circulation of dangerous people. Thus, he regards migration as a central security issue, as illustrated by the title of his essay—"Politics in a World of Hybrid Cultures: Migration is the Central Issue of the 21st Century" —published in 2001. In this essay, Huntington writes:

> Migration is the central issue of our time. In the developed world there is an aging and, soon, shrinking population. In the larger part of the rest of the world a population expansion continues, generating a mainly youthful population that is the source of migration, instability and terrorism. (22)

As illustrated here, youthful populations emerge as the central agents of risk and danger to the security of Western nations.

Melinda Cooper observes that this type of discourse that focuses on the dangers of fecundity and population circulations redefines national security in the broadest of terms. Security, within this new biopolitical discourse, is not limited to the integrity of national boundaries or economies but is broadened to encompass any and all risks potentially threatening "human, biological, and even biospheric existence rather than the formal institutions of the state" (Cooper 93). Fertility, health, movement, and education are all formulated within calculi of risk posed to the socioeconomic stability of (privileged) developed nations. Security today thus embraces nearly all phenomena impacting populations, particularly youthful populations. This new discourse, Cooper argues, renders most all human concerns essential military problems, particularly humanitarian and environmental disasters. She explains that international relations experts are now formally asserting that the major refugee flows of the twenty-first century will "be caused by environmental crisis and conflicts over resource scarcity"; thus, the entire field of humanitarian relief and disaster response consequently finds "itself invested with 'new-found military and diplomatic implications'" (Cooper 92).

In sum, national security has thus become framed as embracing the totality of biopolitical representations of population risks. Humanitarian relief has become inflected with military-security problem-solution frames, and military strategists are increasingly agitating for the adoption of a more pastoral approach toward U.S. military interventions abroad. As will be elaborated upon in the next section, children play important roles in the new security discourse and policy orientations. Depictions of foreign children found in the media mobilize public support in Western nations for new security agendas that extend well beyond protection of national borders. Children also serve as sites for "tutelary" governmental approaches toward managing risky populations. Finally, "ungovernable" children are represented as inviting severe disciplinary action, thereby justifying "police action" interventions by powerful nations.

Western governments can promote their security objectives by shaping media, humanitarian, and philanthropic agendas through news releases and policy statements. Engaging philanthropic support for foreign policy objectives is a tactical strategy that can shift costs away from government apparatuses. Additionally, although Western humanitarian and philanthropic relief efforts take many forms, they increasingly employ market-based technologies such as microenterprise that are consistent with Western market and political logics (Edwards 1–92). Privatizing security in this fashion promotes public support at home and potentially reduces resistance abroad to the implementation of Western models of economic and social development since the "assistance" is represented as humanitarian or philanthropic in intent.

Within the United States, domestic support for official foreign aid lags, given domestic poverty and infrastructure needs. Nongovernmental organizations (NGOs) and various religious organizations play a more important role in delivering aid as official U.S. government funding declines (Vogel 635–55). Children are used by charitable organizations and government agencies alike to build support for, and contributions to, foreign aid programs. Representations of children are important devices for governments and societies aiming to privatize security operations.

Politics infuses the selection of images of children found in Western media and aid appeals (Burman "Innocents" 238–53). The homogeneous frames of Western media depict the children of some regions as deserving victims, whereas the children of other, equally impoverished regions are ignored or rendered invisible. Further, depictions of poor children in mainstream media and NGO appeals usually depoliticize accounts of the children's suffering, offering sympathetic audiences scant understanding of the political and economic factors behind their plight. Western audiences typically pay little attention to relief strategies but rather focus their attention on the innocents' suffering and the tantalizing hope of their redemption through donor generosity. Depictions of children occasionally stray from images of suffering to examine risky or dangerous children whose debasement stems from their indoctrination within terrorist organizations. While child-welfare advocates argue for the rehabilitation of children who have willingly or unwillingly served as combatants, images of children as warriors and terrorists have the capacity to harden popular sentiment against children rendered *other* by virtue of their ethnicity, religion, or geographic locale. The othering of children engaged in conflict may explain why the U.S. public has tolerated the incarceration of children in the name of the "war on terror." Finally, children can assume dangerousness through Western media depictions that emphasize their seeming "overabundance" in poor and developing nations. These depictions emphasize their consumption of resources and link their circulation to the insecurity of privileged populations. The following sections explore these representations in further detail.

Representing Children: Innocent Victims or Dangerous Criminals?

Lacking direct contact with those who live far away, most people in the Western industrialized world rely on a variety of media forms for information about life outside their immediate experiences. Depictions of poor children of the world in Western media oscillate between those of (1) innocent victims of third world disasters and wars or (2) surplus and dangerous populations threatening global stability. Each of these representations is examined.

As Erica Burman points out, children are used extensively in aid appeals targeted at Western, industrialized audiences ("Innocents" 241). Representations of poor children often depict starving infants and toddlers with hollow faces and sunken eyes. Starving infants are seen cradled in mothers' arms or under care workers' nurturance. The children's plight is typically explained in vague terms, as deriving from the anonymous forces of famine or war. Burman claims that these representations can invoke Northern colonial paternalism while infantilizing the South through the images of helpless and hapless children (241). Suffering is decontexualized from its sociohistorical and political situatedness as the camera or news text personalizes and sensationalizes tragic events by focusing on individual experiences. Nicola Ansell elaborates: "Children are represented as innocent, which implies someone must be guilty, and with no evidence that they are supported by family or community, their situation is attributed to the failure of their people to care for them" (28). Northern viewers consuming the texts of suffering children are invited to step in and "save" the poor child from the developing world's incompetence and cruelty.

Burman suggests that these images simultaneously produce a dynamic of otherness and one of identification. While otherness is grounded in the viewers' cultural alienation from the spectacle of difference and suffering, identification occurs through a "variety of cultural chauvinism which also follows from the imperial legacy" ("Innocents" 241). With identification, the viewer assumes a commonality of experience and location that denies real differences and that can therefore operate as a technology of power by rendering invisible Northern privilege. By assuming the universalized role of parent/savior, the Northern viewer participates in the construction of those in the developing world as childlike victims of their own incompetence, thereby inviting the Northern viewer's intercession. Lack of reflexivity about cultural and historical complexities can cause Northern "saviors" to promote or apply "inappropriately homogenised or culturally chauvinistic developmental models" (243) that reinforce global or regional power hierarchies and that potentially disrupt existing modes of life and subsistence patterns. The IMF, the World Bank, and the U.S. Agency for International Development (USAID) have been severely criticized for implementing and occasionally enforcing application of Western development models that are impractical or otherwise poorly conceived for local populations in the developing world. More recently, Western-supported microenterprise and microfinance have eclipsed other assistance models despite criticism that these approaches are insufficient for bootstrapping entire populations out of poverty (see Crewe and Harrison 1–214; Elyachar 1–279).

Interestingly, the media effects described by Burman also apply to the response of some affluent white populations to the images of Hurricane Katrina victims in 2005 (Nadesan "Hurricane" 76–78). In the immediate aftermath of the hurricane, the U.S. news media broadcasted horrific images and narratives of those left behind, including the story of a brutal rape and murder of a seven-year-old girl inside the city's convention center and of snipers shooting medical crews attempting to aid victims (see Garfield 55–74). Images of brutal, lawless, and armed young African Americans mesmerized viewers across the nation. Viewed at the height of neoconservative and neoliberal hegemony, the authenticity of these narratives and images was not questioned. Rather, they serviced neoconservative and neoliberal policy reforms aimed at privatization and segregation because the images purportedly demonstrated the failures of welfare logics and policies and the degradation of New Orleans's poor (Garfield 55–74; Nadesan "Hurricane" 82–83). The result was that New Orleans emerged a whiter, richer, and less populous city than before (Harden "A City's" A1) as the poor were relocated, many in refugee camps of contaminated trailers. Dislocated children of Katrina have failed to receive needed social welfare services:

> After more than three years of nomadic uncertainty, many of the children of Hurricane Katrina are behind in school, acting out and suffering from extraordinarily high rates of illness and mental health problems. Their parents, many still anxious or depressed themselves, are struggling to keep the lights on and the refrigerator stocked. (Dewan A1)

These children's anxieties surely stem from the destruction of their previous neighborhoods and schools as reconstruction has encouraged gentrification and replacement of public schools with charter and private schools. The media depictions of New Orleans's entrenched poverty helped legitimize these reforms, which have together led to a whiter and richer New Orleans and the dispossession of many of its former citizens.

International and domestic images of suffering or victimized children in the media oscillate between representing these children as at risk on the one hand and as risky on the other. Indeed, a number of observers of childhood point out that formulations of children as "at risk" often coincide with conceptualizations of the children as themselves risky. Victimized and suffering children are symbolically "out of place" in the Western imagination and are therefore potentially ominous or threatening. Representations of children as out of place can lead viewers to regard them as posing future risks (Vanobbergen 161–76). Just as nineteenth-century child advocates warned of the contagion of adverse environments, so too do contemporary commentators point to the potential risks posed by children perceived as out of context. Potentially ungovernable, these children invite policies of containment, normalization, and discipline. Worse, they can be perceived as vaguely parasitic or threatening, thereby inviting authoritarian technologies of control, or even elimination. Thus, Sharon Stephens notes that the poor, dispossessed children of the third world are often represented as surplus populations responsible for exhausting global resources:

> Just as idle bands of youth and street children are seen as the cause of urban crime and decay, so also are hordes of hungry Third World children often seen as a major *cause* of global environmental problems...The solution to global environmental problems, such perspectives suggest, is population control programs aimed at drastically reducing "excess populations"— people filling spaces and making demands on natural resources outside socially legitimated channels of ownership, exchange, and distribution. It is significant that in the popular media, these excess populations are often represented by crowds of hungry *children*, consuming whatever resources are available, but supposedly unable to participate in the global economy as producers. (Stephens 13)

Children of the developing world seem in the Western imagination to exhaust global resources despite empirical evidence that Western children consume far, far more in resources.

An example of this misrepresentation of the depletion of global resources is found in a *Wall Street Journal* article on the revival of "Malthusian fears" that emphasizes population growth and spreading prosperity in India and China:

> As the world grows more populous — the United Nations projects eight billion people by 2025, up from 6.6 billion today —it also is growing more prosperous. The average person is consuming more food, water, metal and power. Growing numbers of China's 1.3 billion people and India's 1.1 billion are stepping up to the middle class, adopting the high-protein diets, gasoline-fueled transport and electric gadgets that developed nations enjoy. (*Lahart, Barta, and Batson A1*)

The article fails to mention Western consumption patterns in the United States, Europe, and Japan other than to state that these regions "have proven adept at adjusting to resource constraints" in the past. The world's poorest 20 percent of the population consume approximately 1.5 percent of resources, while the middle 60 percent consume 21.9 percent of resources, and the world's richest 20 percent consume 76 percent of global resources, according to the World Bank Development Indicators 2008. Collection and mapping of biopolitical statistics on resource consumption serve the political purpose of dramatizing gross global inequities. Yet the popular Western imagination is rarely confronted with images challenging its own excess. Instead, in the Western imagination, it is the children of the developing world who present the Malthusian challenge.

For many observers and tourists, the Malthusian challenge of excess children is demonstrated empirically by street children. Just as nineteenth-century street children constituted an unwelcome presence in public spaces, so also do contemporary street children. Amnesty International estimates that there are between 100 million and 150 million "street children" in the world (http://www.amnesty.org/en/children). Street children are commonly treated as if they were invisible in the developing world by locals and visitors alike.[8] Indeed, homeless people in the developing world often face hostility, suspicion, and apathy and tend to be perceived as responsible for their circumstances (Speak and Tipple 172–88). A recent visit to Indonesia caused marked concern among my own children, who, in contrast to the surrounding adults, were struck by the stark visibility of children begging on the streets. Yet adult Asian and Western tourists alike seem to find these children utterly invisible.

Street children have long threatened social sensibilities, but today they also are cast as risks to a globalized social order, particularly through their circulation. The world's poor populations are in motion. The World Bank estimates that there are 74 million "south to south" migrants who move from one developing country to another (DeParle "A Global Trek"). Pushed by wars and famines or pulled by jobs or better wages, these migrant populations search for a better life. Nearly half of these migrants are children. Migrants' children are often denied citizenship in the countries into which they are born. A document promulgated by the Office of the United Nations High Commissioner for Refugees entitled "Refugee Children: Guidelines and Care" suggests how legal treatment may be provided to refugee children, but its guidelines are often ignored or impossible to implement (see United Nations "Refugee Children"). Consequently, the children of migrants often experience abject poverty.

Lacking access to adequate food and sanitation, poor migrant children fail to conform to ideals of health and cleanliness, encouraging perceptions that they are diseased. Thus, their victimization contributes to further victimization as resident populations and visiting tourists alike regard migrant street children unconsciously, or even overtly, as disease-carrying, delinquent-inclined, risky, dangerous persons. As witnessed by my family, the three-year-old child begging on the street in Yogyakarta, Indonesia, is not regarded by pedestrians as innocent or deserving of sentimental pity; pedestrians ignored his empty cup, which happened, ironically, to be a cast-off, paper McDonald's coffee cup.

Risky Children and the Erosion of Pastoral Biopolitics

"Surplus" children of the developing world are increasingly represented as posing dangers beyond their mere circulation and fecundity. Recently the media have been filled with stories of children and youth "at risk" of becoming terrorists. For instance, a 2008 *New York Times* article describes a "Tug of War for Young Minds" in Algiers:

> First, Abdel Malek Outas's teachers taught him to write math equations in Arabic, and embrace Islam and the Arab world. Then they told him to write in Latin letters that are no longer branded unpatriotic, and open his mind to the West.
>
> Malek is 19, and he is confused. "When we were in middle school we studied only in Arabic," he said. "When we went to high school, they changed the program, and a lot is in French. Sometimes, we don't even understand what we are writing."
>
> The confusion has bled off the pages of his math book and deep into his life. One moment, he is rapping; another, he recounts how he flirted with terrorism, agreeing two years ago to go with a recruiter to kill apostates in the name of jihad. (Slackman)

The article describes Algeria's youth as "in play" in the context of a supposed struggle between Muslim religious revival and a secular government effort to resist religious extremism.

A similar article published in 2008 in the *New York Times* describes boys purportedly training to fight U.S. soldiers in Iraq:

> The children in black—T-shirts, trousers and face masks—hoist AK-47s and pistols and rush toward an apparently unarmed man on a bicycle. In an instant they have surrounded him, shouting in the high voices of boys who are not yet men, "Put your hands behind your back.…" The video, shown Wednesday by the military at a news conference, is believed to be part of a propaganda tape made by Al Qaeda in Mesopotamia, the homegrown extremist group that American intelligence officials say has foreign leadership. (Rubin)

Finally, yet another *New York Times* article describes "Hezbollah Shrine to Terrorist Suspect Enthralls Lebanese Children" (Worth). This article describes the seduction of children by terrorists causes, thereby narrating the ways whereby terrorism contaminates childhood innocence.

Sharon Stephens describes how these types of representational practices replace childhood innocence with menace in her account of children's dehumanization:

> In the darkest scenarios of the disappearance of childhood, the theme of lost innocence takes on a negative valence. Children on the streets of Rio de Janeiro or in the ghettos of Los Angeles are not only killed. They also kill. Unrestrained and undeveloped by the ameliorating institutions of childhood, the innocence of childhood is perverted and twisted. In these stories, children are represented as malicious predators, the embodiment of dangerous natural forces, unharnessed to social ends. (13)

Stephens points out that what is so ominous about these types of accounts is that they allow acts of violence against children to occur that would otherwise be considered unconscionable. Once children are viewed as "*outside* the normative socializing control of adult society," as "unsocialized or antisocial dangers to the established order and as primary *causes* of escalating social problems," it becomes possible for police and other authorities to engage in acts of brutality against children. The use of death squads against Brazilian street children illustrates this possibility (11–12). It also becomes possible to render these children utterly invisible, to shut out the misery of their hunger, sickness, and destitution.

Some of the most horrific examples of contemporary Western suspicion and victimization of children stem from the U.S. war on terror and occupation of Iraq. In particular, the loss of childhood innocence is revealed in the U.S. treatment of children as potential enemy combatants. The U.S. government has imprisoned over 2,500 children in Guantanamo since 2001 as "enemy combatants." Some of these children were as young as twelve years old when they were incarcerated (see Glaberson). Xenophobic attitudes about foreign terrorists transform understandings of childhood, erasing any sense of childhood innocence and vulnerability and precluding any form of special protections.

This policy of treating children as enemy combatants has tragic consequences, as the following passage shows:

> In the 2004 assault by US Marines on the city of Fallujah, things were even worse. Dexter Filkins, a reporter for the *New York Times*, reported that before that invasion, some 20,000 Marines encircled the doomed city, which the White House had decided to level because it harbored a bunch of insurgents and had angered the American public by capturing, killing and mutilating the bodies of four mercenaries working for US forces. The residents of the 300,000-population city were warned of the coming all-out attack. Women and children and old people were allowed to flee the city and pass through the cordon of troops. But Filkins reported that males determined to be "of combat age," which in this case was established as *12 and up*, were barred from leaving, and sent back into the city to await their fate. Young boys were ripped from their screaming mothers and sent trudging back to the city to face death.

In the ensuing slaughter, as the US dumped bombs, napalm, phosphorus, anti-personnel fragmentation weapons and an unimaginable quantity of machine gun and small arms fire on the city, it is clear that many of those young boys died. (Lindorff)

Although Lindorff's claim that boys were turned back by U.S. soldiers cannot be authenticated with certainty, the record is clear that the United States deliberately used phosphorus against human populations, including children, in the attack on Fallujah.[9] In 2009 the United States was accused of using white phosphorus against civilian populations in Afghanistan after Afghan doctors found patients with severe and unusual burns (Straziuso and Faiez A4). Unmanned drones used to combat "terrorists" in Afghanistan have also produced large numbers of child casualties.

In 2007 UNICEF asserted that hundreds of Iraqi children were killed in violence and that 1,350 were detained by authorities (cited in "Iraq Children"). Many more, upward of 2 million, face poor nutrition, lack of education, disease, and violence in Iraq because of the war unleashed by the U.S. invasion. Yet there are fewer stories in the mainstream U.S. press covering the plight of Iraqi children than there are stories addressing the purported loss of childhood innocence in the region. The media are captivated by the spectacle of dangerous children. Children in these accounts are represented as incipient terrorists and potential threats to U.S. security as their innocence is subjected to lethal perversion.

The dehumanization of children as enemy combatants also occurred recently in the context of Israel's bombing of Gaza. Half the population of Gaza is under the age of sixteen years. Gaza children, already facing malnutrition as a result of the Israeli embargo, were subject to Israel's January 2009 retaliation against Hamas. More than 400 children were killed in the Israeli assault, some by bullets and some by air assaults (Fraser). The children who survived are traumatized:

> Even the children who escaped physical injury face the psychological consequences of having lived under near-constant bombardment for 22 days and nights. A week into a fragile cease-fire, mental health experts, human rights advocates and parents say they worry that this generation of Palestinian children will suffer the effects of the war for decades to come. (Witte A5)

Israel's disproportionate retaliation against Hamas produced a spectacle of death and misery that led the pope to describe Gaza as a "big concentration camp" ("Vatican Deplores"). Israel's response that it had the right to defend itself against terrorist attacks had the practical effect of erasing any distinctions among Gaza's inhabitants: all were cast as potential enemy combatants. By February 2009, two-thirds of Gaza's residents lacked power, and a third lacked running water; nearly all lacked basic supplies and faced overwhelmed medical facilities ("Gaza Humanitarian"). In June 2009 the former U.S. president Jimmy Carter claimed, "Never before in history has a large community been savaged by bombs and missiles and then deprived of the means to repair itself" (cited in Schneider A1). Israel's actions have surely been more persuasive than any appeals by Hamas to take up arms against Israel.

Children exposed to, and subjected by, military and guerilla force do sometimes pursue violence. Children's role in the discourse and practices of war has expanded beyond that of innocent victim as more children are conscripted into, or willingly join, conflicts across the globe. Peter Singer, the author of *Children at War*, reports in an online interview that

> children now serve in 40% of the world's armed forces, rebel groups, and terrorist organizations and fight in almost 75% of the world's conflicts; indeed, in the last five years, children have served as soldiers on every continent but Antarctica. An additional half million children serve in armed forces not presently at war. The children are often abducted to fight and participate in all the full horrors of war; indeed they are sometimes forced to carry out atrocities that adults shy away from. (Singer "Young")

A December 2008 *New York Times* video portrays rebel soldiers with youthful faces accused of massacring villagers in eastern Congo ("Mass Killings"). The narrator does not comment on the youth of the rebel soldiers, but mentions that one villager's fifteen-year-old son was kidnapped by the rebels. The conscription of unwilling children entails brute force and insidious indoctrination, often using drugs and alcohol. Policymakers, politicians, and adult military forces grapple with how to differentiate dangerous children from innocent ones. Likewise, policymakers weigh whether to rehabilitate or punish child suspects. In Fallujah, it appears that the boundary separating sympathetic child victims from dangerous child soldiers may have been obscured as U.S. soldiers sought to contain and punish the city's recalcitrant population.

The strategic use of children in African nations to kill civilians demonstrates the loss of childhood innocence and reinforces Western perceptions of regional barbarism. However, these representations obscure the historical legacies of Western colonialism and the contemporary international community's complicity in enabling civil conflict and strife through the manufacturing and large-scale distribution of small arms and through the militarization of aid in Africa. After September 11, 2001, the U.S. Pentagon concluded that Africa's impoverished populations and "ungoverned spaces and ill governed states" posed a growing risk to U.S. security, precipitating the militarization of U.S. aid to the resource-rich continent (Shanker "U.S. is Top," "Weapons Sales"; see also McCrummen A10). The U.S. Africa Command (AFRICOM), which became operational in 2008, was instituted by the former U.S. president G. W. Bush as a U.S. military central command for the continent (Engdahl 227).

U.S. military aid collapses the distinction between humanitarian support and militarization, as illustrated in this passage explaining how aid will be implemented:

> Mr. Gates and General Ward said that this work to complement and support American security and development policies would include missions like

deploying military trainers to improve the abilities of local counterterrorism forces, assigning military engineers to help dig wells and build sewers, and sending in military doctors to inoculate the local population against diseases. (Shanker "Command for Africa")

Vaccination and training of (para?)military forces achieve equal footing within this approach to security. McCrummen reports in the *Washington Post* that the growth of military aid has occurred in tandem with the decline in aid for international development: "The Pentagon, which controlled about 3 percent of official aid money a decade ago, now controls 22 percent, while the U.S. Agency for International Development's share has declined from 65 percent to 40 percent" (A10).

Superpowers' interests in the Democratic Republic of the Congo's (DRC's) resources have contributed to the civil unrest in the nation. Over 5,400,000 Congolese civilians have died in conflict since 1996 (Engdahl 27). This conflict, which is often represented in terms of ethnic tensions, has been fueled by armed sales and training to various groups within the country and by incursions or interference by the neighboring nation of Rwanda. For instance, McCrummen reports that $5.5 million in aid was delivered to Congo by the United States to boost and professionalize an army of 164,000 soldiers. Congo's strategic importance stems in large part from its possession of Coltan, a vital metal necessary for mobile phones (Engdahl 230). Children caught up in the Congolese conflicts may rightfully be considered pawns in resource wars between and among petty sovereigns and superpowers. The debasement of childhood innocence in media representations of child soldiers helps legitimize more U.S. aid to, and policing of, the country. Engdahl calls this type of rationalization "weaponizing human rights" (89).

The U.S. government and arms industries play important roles in global weapons proliferation that endangers civilian populations worldwide (see C. Johnson 1–367). In 2005–2007 the United States dominated both global weapons sales and sales to the developing world, followed by Russia, Great Britain, and China (see Shanker "U.S. is Top," "Weapons Sales"). China plays an important role in proliferating smaller weapons. Major U.S. defense contractors also play a role in the global proliferation of weapons, including Northrop Grumman, General Dynamics, and Lockheed Martin. The wars in Iraq and Afghanistan have reportedly created a financial bonanza for these contractors (Merle "Defense Earnings" D1), but they are merely the tip of the iceberg of entire industries aimed at the creation and circulation of weapons. Small and light arms are particularly likely to be used against civilians, and yet the U.S. stance has been to condemn illegal trafficking while failing to recognize the link between legal and illegal trafficking (Lumpe). In 2001 the U.S. government blocked efforts by the UN to control small arms sales, claiming that the right to bear arms was an issue of personal freedom. A U.S. official, John Bolton, claimed that a distinction existed between firearms used for traditional and cultural reasons, and firearms traded illegally around the world ("U.S. Blocks"). From 2001 U.S. arms sales tripled, growing to $32 billion in 2007 (Schweid). U.S. policy has disregarded the intimate connection between crime, violence, and the circulation of arms and torture devices (Nordstrom 93–95).

The U.S. disregard for international efforts to combat proliferation of light arms is part of a more general pattern of disregard for human rights outside of specifically targeted concerns. For instance, the United States has specifically addressed the sexual trafficking of children and human rights abuses abroad against select Christian populations because of religious constituents within the United States In contrast, the United States has demonstrated blatant disregard for international efforts to combat the detention and punishment of children who are implicated by, or in, adult conflicts. The practices of detaining, incarcerating, targeting, and punishing children for adult conflicts contradict the principles of child protection codified in the United Nations Convention on the Rights of the Child (General Assembly resolution 44/25 of November 20, 1989). Entered into force on September 2, 1990, Article 37 of the convention states:

> No child shall be subjected to torture or other cruel, inhuman or degrading treatment or punishment. Neither capital punishment nor life imprisonment without possibility of release shall be imposed for offences committed by persons below eighteen years of age; No child shall be deprived of his liberty unlawfully or arbitrarily. The arrest, detention or imprisonment of a child shall be in conformity with the law and shall be used only as a measure of last resort and for the shortest appropriate period of time.... (http://www2.ohchr.org/english/law/crc.htm)

Ratifying states are legally bound by the convention's stipulations (Stephens 35). Although often applauded, the convention has not been without criticism. Some critics argue that the convention tends to normalize and elevate Western, paternalistic constructions of the family (John 107), while others question the value or utility of efforts to create globalized standards and universalized policies for protecting children (Boyden 208). Despite these criticisms, the convention offers children some legal protection against precisely those forms of treatment perpetrated by U.S. officials and military personnel during the occupation of Iraq and the war on terror.

In sum, representations and materializations of suffering children often lead observers to promote policing or containment. Western policy analysts see excess populations of poor children as resource depleting and potentially destabilizing. However, while children's circulation and migration are seen as threatening political and social orders, children can also serve as useful pawns in superpower politics within geopolitically strategic regions. Their relative pliability allows them to be easily conscripted for resource-driven conflicts. Representations of child soldiers in Western media legitimize humanitarian interventions and policing by superpowers seeking to gain official access to a region. Additionally, Western media–promulgated images of child "terrorists" quell global dissent over the brutal and forceful treatment of children aligned with opposing forces. Sovereign strategies of force—including incarceration, enforced starvation, and offensive military action—disrespect international law and humanistic governmental logics that have sought to protect children over the last sixty years.

Poor refugee children who end up in camps or foreign cities are driven by many forces, including civil conflict, impoverishing government policies, and drought and deforestation. Refugee children are subject to dehumanization by local populations who view them as pre-delinquents or as vectors of contagion. Local urban populations, especially those who cater to tourist trades, may argue for these children's containment in enclosures. Sometimes they are even targeted for extermination, as has been the case with Brazilian street children (Stephens 12). Western consumers of images of poor, suffering children abroad often interpret the children's plight in relation to colonial-tinged world views, which reinforce Western superiority while inviting Western intercession. At the same time, Western audiences are rarely interested in encouraging large numbers of child refugees to immigrate to their nations. The next section addresses the circulation of children in the context of immigration anxieties.

CHILDHOOD INNOCENCE, CONTAINMENT, AND POPULATION FLOWS IN THE UNITED STATES

Illegal immigration has increased significantly in the United States within the last twenty years, particularly from Mexico. Media accounts and the sheer visibility of immigrant children within industrialized nations have encouraged strong anti-immigrant sentiment. Children emerge within anti-immigrant rhetorics as fundamentally threatening to national security, which is defined within nativistic and racist constitutions of national identity. This section of the chapter explores representations of immigration and efforts to govern immigration flows using sovereign and disciplinary technologies. It examines how xenophobic discourses of immigration contribute to further unraveling of social welfare entitlements within the United States, exacerbating individual risks in a time of economic uncertainty while also encouraging authoritarian treatment of "risky" populations—illegal immigrants and their citizen children, unemployed youths, and others—who are perceived as public security risks or burdens.

NAFTA, or the North American Free Trade Agreement, is a neoliberal trade agreement signed into law by President Clinton in 1994. NAFTA's populist detractors warned of risks to American manufacturing, but perhaps those most burdened by the "risks" of free trade were Mexico's small farmers, who were unable to compete against U.S. agribusiness, particularly U.S. corn (Stiglitz "The Broken"). Mexican small businesses and manufacturers were also driven out of business by the new competition (Uchitelle). Unable to survive under the new NAFTA-created conditions, many Mexicans migrated to Mexico's megacities and, eventually, to the United States. The Pew Hispanic Center reports that two-thirds of undocumented immigrants currently living in the United States have lived there for ten years or less (cited in

Dickerson C1). In 2005 the *New York Times* estimated that 5 million Mexican citizens were living undocumented in the United States (DePalma).

The total number of Mexican citizens living illegally in the United States is actually a relatively small number compared with the overall U.S. population. However, popular misperceptions about the number of undocumented immigrants may stem from a racist backlash against growing diversity within the U.S. population. That population segment that defines itself as Hispanic is clearly growing because of its relative fertility and, to a lesser extent, immigration. One in four children under five years of age is categorized as Hispanic in the United States This population grew to 15.1 percent (at 45.5 million) of the total U.S. population in 2007. African Americans are the second-largest minority, with a population of 40.7 million. Minorities overall accounted for 34 percent of the U.S. population. The Hispanic population is expected to double to 30 percent of the U.S. population by 2050 (Aizenman "1 in 4" A3).

As established in previous chapters, minority populations disproportionately experience poverty within the United States. Recent immigrants are particularly likely to work in low-paying jobs lacking benefits and security. Many immigrants from poor countries often lack the formal education and skills necessary for upward mobility in an increasingly polarized U.S. economy. Consequently, their children may also have little access to the material and symbolic capital necessary for social mobility as they are forced to attend the underfunded and crowded schools described in previous chapters. Immigrants specifically and poor populations more generally are at increasing risk of becoming a permanent underclass.

Social mobility has eroded within the United States because of automation, globalized production, and the particular characteristics of a service economy whose living-wage jobs require a college degree at least and often require advanced, technical education and experience (see Henderson D1). Illegal immigration may put downward pressure on particular job sectors such as agriculture and low-skilled manufacturing, but only because of the willingness of industries to rely on workers whose fear of reprisals undermines their ability to demand living wages and benefits. However, downward pressure on work conditions exists in occupations not populated by immigrants. White, working-class populations are also vulnerable to economic exploitation by unscrupulous employers. The poor wages, lack of full-time work, and lack of benefits associated with retail work illustrate the white working-class population's economic vulnerability in the absence of competition from immigrant workers.

Working-class populations who fail to achieve the successes associated with the "American Dream" are often depicted as responsible for their own plight. Impoverished immigrant populations, legal or otherwise, receive little sympathy from a public that has next to no historical memory of how labor gains were achieved over the last one hundred years. Past economic exploitation and vilification of early twentieth-century European immigrants have been erased from the public mind, these memory traces having been replaced by a narrative of upward mobility achieved by industrious, market-savvy entrepreneurs. The narrative of individual striving and success has no space for labor agitation, strikes, and confrontation with exploitative capitalists. And so the contemporary public imagination conceives of present immigrants as less industrious, less skilled, and less worthy than those of past waves.

A curious form of *ressentiment* held by middle-class Americans facing job insecurity and wage pressures further undercuts the power of working-class populations to demand better work conditions while, simultaneously, exacerbating xenophobic tendencies against immigrants. Ressentiment, as expressed by Friedrich Nietzsche (1844–1900) in his *Genealogy of Morals*, refers to how one group's sense of inferiority is projected onto a scapegoat. White-collar, middle-class populations have long differentiated their status from working-class populations by the (former's) job security and the symbolic capital associated with job autonomy and annualized compensation with benefits. The erosion of middle-class wages, benefits, and security has been coupled with assaults on worker autonomy through new workplace technologies of accountability and visibility. Even highly skilled workers such as college professors, lawyers, and medical doctors experience less job security and autonomy than previously enjoyed. Middle- and upper-middle-class anxiety rarely confronts employer and industry practices, but rather fixates instead on the supposed excesses of those dwindling working-class workers protected by unions (e.g., autoworkers, school teachers, and UPS delivery persons) as well as the supposed menace posed by poor immigrants.

Unionized blue-collar workers, poor immigrants, and minority populations emerge in a politics of ressentiment as responsible for the erosion of the American dream. This politics of ressentiment is clearly evident in the popular blame attached to worker pensions for the collapse of the U.S. auto industry, erasing years of deliberate underfunding of pensions by the auto industry executives (Lowenstein A1). It is also evident in increasingly vociferous attacks against immigrants for undermining wages, national security, health care, and the nation's educational achievements, as measured by standardized test scores.

Immigrants, especially Hispanic immigrants, have been subjected to brutal attacks and treatment as a result of growing anti-immigrant sentiment. In 2008 a poor immigrant from Ecuador was killed as part of a spree by teenagers who tormented immigrants with knives and BB guns ("Police"). In 2008 a Mexican family's home in the same geographic area was burned, and in 2001 two Mexican day laborers were nearly beaten to death (Semple A25). These chilling examples illustrate the excesses of the politics of ressentiment.

Poor treatment of immigrants also occurs at the hands of security officials tasked with "protecting" the homeland. The U.S. National Network for Immigrant and Refugee Rights argues in a report titled "Human Rights and Human Security at Risk" that abusive and discriminatory immigration enforcement was exacerbated by the relocation of immigration enforcement within the Department of Homeland Security (DHS) in 2003 (Nimr,

Tactaquin, and Garcia). The U.S. Government Accounting Office (GAO) issued a report in 2007 detailing that the number of immigrants detained between 2002 and 2007 grew from 90,000 to 283,000. The report claims that detainees were often improperly barred from making a single phone call to a lawyer (cited in Hsu "No Phone" A2). In February 2009 the DHS inspector general Richard Skinner reported publicly that 108,000 deportees had children who were legally U.S. citizens (Gamboa A21). Failure to explain what happened to the citizen children of deported parents raises the specters of forcibly abandoned children and deportations of child citizens.

The U.S. government treats illegal immigrants as criminals, although most have not committed any crimes beyond their illegal entry, and many detainees have committed only minor administrative violations such as over-staying their visas ("Death"). As reported by Hsu in the *Washington Post,*

> Federal law enforcement agencies have increased criminal prosecutions of immigration violators to record levels, in part by filing minor charges against virtually every person caught illegally crossing some stretches of the U.S.-Mexico border, according to new U.S. data. (Hsu "Immigration Prosecution" A1)

In May 2008 a raid of a meatpacking plant in Iowa resulted in the criminal prosecution of 297 illegal immigrants. The decision to sentence the immigrants under federal criminal law, rather than civil statutes, led to five months of imprisonment followed by immediate deportation (Weintraub).

Growing numbers of immigrants are subjected to incarceration before deportation. Paromita Shah, the associate director of the National Immigration Project of the National Lawyers Guild, argued in 2008 that "Homeland Security is one of the largest jailers in the world, 'but it behaves like a lawless local sheriff'" (cited in N. Bernstein B3). Since 2004, eighty-three immigrants have died in retention centers: critics claim that detainees are routinely denied basic rights, such as medical care, visitation, and legal materials (N. Bernstein B3; Bernstein and Preston A18; Gorman B5). The DHS inspector general reported in 2006 that an audit of five publicly and privately operated detention centers found all of them out of compliance with general standards on health care, disciplinary procedures, and access to legal materials. Inexplicably, the five had been rated "acceptable" in the immigration agency's annual reviews (cited in N. Bernstein B3). A separate report criticizing practices of immigrant detention issued in 2007 noted that at the privately run Hutto detention center in Texas, women received inadequate prenatal care, children received only one hour of schooling daily, children as young as six years of age were separated from their parents, and threats of separation were used as "disciplinary tools" (Swarns A17).

Detention facilities have followed the trajectory set by the outsourcing of U.S. prisons. The Corrections Corporation of America and the GEO Group have successfully contracted with the federal government to provide immigration detention services (Gorman B5). The federal government also contracts with state and local jails and criminal detention centers to hold immigrant detainees. Private contractors are not subject to the same freedom of information (FOI) provisions applicable to government-run agencies, and private criminal detention centers treat immigrants as if they were criminals. Consequently, a GAO report concludes, "Without sufficient internal control policies and procedures in place, ICE [Immigration and Customs and Enforcement] is unable to offer assurance that detainees can access legal services, file external grievances and obtain assistance from their consulates" (cited in Hsu "No Phone" A2). Immigrants continue to be denied basic rights and due process because of the lack of universal, enforceable standards governing detention (N. Bernstein B3).

Public sentiment increasingly allows this type of sovereign intervention and incarceration despite the protests of immigration and religious groups and despite a majority consensus among Americans that immigrants are hard working and create jobs (Constable A15). Public sentiment in the United States and Europe is particularly hostile toward Muslim immigrants: only 54 percent of American and 47 percent of European survey respondents reported believing that Western and Islamic ways of life are reconcilable (Constable A15). Dispersed anxieties about cultural encroachments and job loss manifest themselves in new disciplinary and security apparatuses aimed at securing everyday life from the various threats believed to be posed by people who are perceived as culturally different and economically marginal.

The cultural backlash against immigration has contributed to a seemingly growing lack of tolerance for individuals who fail to conform to normative expectations or who demonstrate reduced capacities for self-governance (see Rose *Politics*). Concerns about public health and safety are increasingly viewed as countering, or even outweighing, individual liberal rights, even when individuals bear little personal responsibility for their (perceived) deficiencies. Nikolas Rose argues that an epidemiological model of preventive public health increasingly frames public attitudes toward those individuals regarded as potentially dangerous because of their marginal biopolitical status: AIDS patients, juvenile delinquents, gang members, welfare mothers, promiscuous women, illegal immigrants, and alien others threaten the economic and social security of the nation. Their deviant conduct and perceived incapacities for self-care are viewed as warranting greater surveillance and disciplinary effects.

Upper- and upper-middle-class white populations have responded to perceived threats of dangerous people by retreating into distant suburbs and gated communities. Surveillance cameras are increasingly located in public spaces such as parks, public streets, and malls to protect against crime and dangerous persons. Ben Chappell developed the idea of "threat governmentality" to refer to this surveillance of space and population in the context of a panopticon of threat (314). Individuals perceived as ethnic minorities are disproportionately caught up in the panopticon of surveillance extended and legitimized by the Patriot Act and other homeland security initiatives. Importantly, Chappell observes, liberal notions of citizen rights tend to be outweighed by the epidemiological construction of public security risk while empowering "petty sovereigns" tasked with protecting public security.

Militia groups have formed among more radical right-wing elements of the population. These petty sovereigns pledge to police the border between the United States and Mexico in order to stem the tide of immigrants. The freeways and highways adjacent to the U.S. border in Arizona and California have become security zones policed by armies of public, privatized, and self-deputized (i.e., militia) "security" officers. For instance, a trip between Phoenix, Arizona, and San Diego, California, entails multiple security stops manned by heavily armed state officers, federal agents, and police dogs. Citizens are required to explain their destination and reason for travel. Resistance results in intensive, secondary inspections. Day travel produces images of handcuffed illegal immigrants sitting on the edge of the freeway guarded by armed officers. Night travel reveals hovering police helicopters' nighttime search lights. These images await travelers venturing between the city of Phoenix and San Diego's beaches.[10]

The near financial collapse of the United States that began in the fall of 2007 no doubt will exacerbate xenophobic attitudes toward immigrants and promote greater tolerance of ruthless deportation strategies. The recently appointed DHS director, Janet Napolitano, suggests that the recession offered an opportunity for enhanced immigration enforcement: "The weak economy is giving the government a unique chance to toughen its efforts against illegal immigration, but officials need to act before the window of opportunity closes" (quoted in Levine). Steep job losses in construction have impacted foreign-born workers disproportionately ("Job Losses"). Competition for the remaining construction jobs will predictably amplify anti-immigrant sentiment.

The Pew Hispanic Research Center reported in 2008 that the median annual income of noncitizen immigrant households fell 7.3 percent from 2006 to 2007 (Kochhar). This decline applies to 7 percent of all U.S. households. The Migration Policy Institute expresses concerns in a *Washington Post* article that illegal immigrants unable to return home because of family and economic concerns may be forced to take very poorly compensated jobs with poor work conditions (Aizenman "Recession Unlikely" A8). Immigrant families generally, and illegal immigrant families more specifically, face severe hardships as the recession limits job opportunities and charitable organizations such as food banks face overwhelming demand. Yet economic insecurity is likely to undermine U.S. citizens' sympathy while exacerbating xenophobia.

CONCLUSIONS

In conclusion, globalizing neoliberal policies have produced vast numbers of circulating people who join immense and growing slum cities or migrate to other nations. U.S. security discourses cast these circulating and migrant populations as risky subjects whose biovitality threatens U.S. economic and political interests. Military and development strategies and technologies have merged as security experts seek to govern unruly population flows. Representations of children in the Western media reflect these concerns, projecting ambivalent attitudes about impoverished children. On the one hand, media representations of impoverished and dispossessed children elicit sympathy among the world's privileged as the images emphasize childhood innocence and victimization. Philanthropic aid promises to redress poverty's excesses without disrupting market operations or the fundamental power and resource inequalities existing between philanthropists and recipients. On the other hand, in media representations and security discourses, dispossessed and impoverished children are increasingly cast as risky, even dangerous, subjects. Their purported loss of innocence allows brutal force to be used against them in the context of the (now renamed) war on terror, as well as in ubiquitous regional conflicts.

When migrant flows enter the United States, nostalgic attitudes about childhood innocence are cast aside as xenophobia prevails. Although migrant workers within the United States are welcomed by employers who often economically exploit their vulnerabilities, childhood populations are viewed as threatening national biovitality. Child migrants and the children of recent immigrants face hostility and eroding social welfare investments, including educational ones. Xenophobic attitudes encourage and facilitate criminalization of immigration, leading to the detention of large numbers of people in public and private institutions. In these institutions, unfortunate immigrants are treated like risky criminals and are therefore denied access to basic commodities and liberties. Thus, they exist without the rights afforded citizens. A powerful domestic U.S. politics of ressentiment overwhelms social activists' and religious groups' denunciations of these strategies of containment and punishment.

CHAPTER 6

Children and the Twenty-First Century: Risky Economies

The previous chapters have addressed the biopolitical risks posed to and by children at the turn of the twenty-first century. This chapter concludes this book by exploring these risks in relation to the momentous events surrounding the financial depression that began in 2007. The discussion draws upon ideas and themes developed in previous chapters concerning the government of U.S. children and explores their implications in the context of the economic challenges produced and exacerbated by neoliberal risk frames.

BACKGROUND

The U.S. financial crisis officially began in December 2007. The crisis was purportedly precipitated by the implosion of the subprime mortgage market, but the roots of the crisis can be traced deeper to the shifts in logics of risk discussed in previous chapters, particularly chapter 3. These shifts in logics and technologies of risk management precipitated a crisis that will have profound effects for the near-term government of children in the United States.

One important shift in logic responsible for the crisis is the shift away from the Fordist logic of mass production toward a logic of financialization. Gerald Epstein defines financialization as "the increasing importance of financial markets, financial motives, financial institutions, and financial elites in operation of the economy and its governing institutions" (cited in Vasudevan). The "financialization" of the economy occurred in response to the limits of capital accumulation posed by the economic stagnation characteristic of advanced, monopolistic corporate capitalism (Foster and Magdoff 64–67). Financialization processes over the last thirty years led to a massive expansion of financial services' contribution to the nation's gross domestic product (GDP—a measure of the total value of goods and services produced in the nation) and contributed to a sharp rise in corporate profits for companies that expanded their financial services (e.g. General Electric). Financialization shaped work and everyday life through the expansion of consumer credit and the transformation of social expectations and perceptions. Randy Martin argues that financialization shapes social relationships so that "social affiliations are reconfigured to extract wealth as an ends by means of risk management" (*An Empire* 7).

The financialization of economic production and everyday life produced a cultural economy of credit-fueled consumption by individuals who might otherwise have experienced a decline in living standards due to stagnating or falling wages. Many among the 75 percent of the U.S. population who experienced falling or stagnating wages were distracted from wage losses by perceptions of increased wealth that derived from real-estate gains and phenomenal stock market returns (see R. Matthews A2). As corporations shed defined-benefit pension plans, individuals invested in stocks directly or through 401(k) plans. Stock ownership created a perception of wealth: between 1983 and 1999 the Dow industrial average total compounded returns exceeded 18 percent a year (Milken). Still, wealth was very unevenly distributed; for instance, in 2003, the top 1 percent of households owned 57.5 percent of corporate wealth (Johnston). The gross unevenness was masked because relatively low interest rates and widely available credit enabled consumer spending. Additionally, these factors fueled the housing bubble, raising housing values and thereby allowing many individuals to draw upon their home equity, transforming their houses into ATM machines (Milken).

The subprime market collapse, consisting of the securitization and global dispersion of securities backed by relatively high-risk mortgage loans, precipitated the global depression that began in December 2007. However, mortgage foreclosures based on consumer indebtedness alone do not explain market collapse. Securities were leveraged in complex derivatives insured by credit-default swaps. Governor Frederic S. Mishkin of the Federal Reserve Bank delivered a speech at the U.S. Monetary Forum on February 29, 2008, in which he explained that the U.S. residential-market mortgage meltdown initially led to credit losses of around $400 billion, which constituted less than 2 percent of the outstanding $22 trillion in U.S. equities (Mishkin). The credit market and stock market meltdowns ultimately stemmed from the leveraging of assets into complicated structured products such as collateralized debt obligations. Neoliberals argue that consumer fraud, artificially low interest rates (because of government interference), and the securities' lack of transparency distorted market operations, thereby causing the meltdown. Efforts to "fix" the markets have not tried to alter their fundamental operations but rather have sought to enhance transparency and to increase liquidity through government injections of capital. I will argue that homo oeconomicus currently continues to reign even as the forces of the state are instrumentalized to his ends. This uninterrupted hegemony signals extension of neoliberal logics as the state itself is deployed in the service of financial interests.

The crash resulting from the leveraging of unsound assets will shape the future of American childhood by exacerbating income declines and public sector debt, further eroding residual social welfare logics and apparatuses.

These trends could also contribute to the erosion of the United States' stature as a superpower, potentially disrupting the hegemony of U.S. neoliberal logics of global market government. Dislocation of neoliberal logics, strategies, and technologies could lead to overtly repressive state sovereignty, since U.S. neoliberalism is backed by a variety of repressive security apparatuses and is inhabited by a lurking authoritarian spirit. The new world (dis)order that shows signs of potentially emerging would pose significant challenges for children in the United States and abroad throughout the twenty-first century. Discussion turns now to examine the national and international economic developments that may very well shape the experiences of U.S. childhood over the next thirty years.

The Bailout: Socializing Risks while Privatizing Profits

In the fall of 2008, the U.S. government responded to the financial meltdown with an unprecedented government bailout. Keynesian economic principles promote increased government spending when private spending declines rapidly. Increased government spending substitutes for private spending as the driver of aggregate demand. Ideally, Keynesian government spending should develop national infrastructure so as to facilitate the private sector's recovery and long-term stability. By increasing the national GDP through its investments, government expands the tax base that allows repayment of national debt accrued through deficit spending. The mercantile logic of investing in the health and welfare of the population also plays an important role in Keynesian economics because these investments are believed to foster long-term aggregate demand.

Conservatives and neoliberals reject deficit spending because it is seen as crowding out private borrowers and as increasing inflation. However, the tremendous impact of the meltdown seemed to demand a strong Keynesian fiscal response, drowning out neoliberal and conservative complaints. Now that the details and initial effects of government spending allocations have become clear, many economic critics on the left and right are questioning their appropriateness and adequacy. These concerns are examined briefly in the context of an empirical discussion of the size and forms of assets involved in the meltdown.

In April 2009 the International Monetary Fund (IMF) stated that the global losses resulting from the financial meltdown exceeded $4 trillion. Approximately $2.7 trillion of those losses derived from loans and assets originating in the United States (Landler and Jolly). American investment and commercial banks alike suffered from the collapse of derivative markets mostly linked to the imploding U.S housing market. Housing prices continued to collapse as foreclosures and unemployment rose in 2008 and the first half of 2009, and more and more homeowners saw equity evaporate from their homes, complicating their efforts to refinance mortgages and encouraging individuals to simply walk away from mortgage obligations.

In 2008 the U.S. government launched the $700 billion Troubled Asset Relief Program (TARP), which provided funds to insolvent and distressed banks and financial corporations, including AIG and Fannie Mae ("Cash Machine" 58–59). While it is impossible to ascertain the total costs to the United States posed by the bailouts and fiscal stimulus, one thing is certain: costs keep accruing. President Obama's Term Asset-Backed Securities Loan Facility (TALF) may expand to $1 trillion (Cho and Irwin D1). The U.S. taxpayers' exposure to the troubled insurer of credit-default swaps, AIG, is $163 billion alone as of March 2, 2009. *MoneyNews* places the total cost of the U.S. bailout in November 2008 at $7.7 trillion (Koprowski). Nouriel Roubini estimates that the bailout will add $7 trillion to public debt (Fallows 90). Roubini's *RGE Monitor* report of March 11, 2009, describes the total U.S. government debt at $10.6 trillion as of January 31, 2009 (Roubini "U.S. Government").

The U.S. domestic policy response has been described by neoliberal and conservative critics in terms of "Keynes on steroids" (Pearlstein D1). Some economists claim that massive government deficit spending is warranted given the level of economic decline (see Delong). However, the point of government spending in the long term within a Keynesian framework is to reignite private demand. Deficit spending that fails to stimulate long-term economic activity will be difficult to pay back in future years. Therefore, for Keynesian policies to work effectively, they must engineer the longer-term vitality of national businesses and the populace.

Many critics have expressed concern that the majority of government spending aimed at the recession has *not* been dedicated to policies and programs that will directly revitalize the nation's infrastructure, its businesses, or its populace. Instead, critics argue, most government spending has been targeted at a bailout of fundamentally insolvent financial institutions (e.g., see Johnson 46–56). These "zombie" institutions remain too weak to lend because they struggle to accumulate capital reserves. Simon Johnson, a former chief economist at the IMF, argues that the government bailout priorities and strategies point to a "quiet coup" of the nation by financial institutions (46). Indeed, the disbursement of federal funds outlined in the May 2009 *Atlantic Monthly* reinforces Simon's argument: only $787 billion went to the stimulus spending (i.e., the American Recovery and Reinvestment Act) compared to $3.25 trillion for the bailout (Lavin and Bachman 58–59).

The trillion-dollar bailout of financial institutions has been represented by official U.S. policymakers as necessary to reignite lending, thereby reigniting consumption and production. Critics describe this process as reinflating the bubble and argue that this pseudo-Keynesian response is simply a massive transfer of wealth from the populace at large to a small group of wealth holders. Accordingly, Michael Hudson argues that the financial bailout is "the largest and most inequitable transfer of wealth since the land giveaways to the railroad barons during the Civil War era." The transfer has occurred with the socialization of private losses as government has bailed out financial institutions and backstopped losses. David Einhorn argues in a speech titled "Private Profits and Socialized Risk" that the owners, employees, and creditors of the financial institutions responsible for the meltdown are "rewarded when they succeed, but it is all of us, the taxpayers, who are left on the hook if they fail. This is called private profits and socialized risk. Heads, I win. Tails, you lose. It is a reverse-Robin Hood system" (7). Joseph Palermo describes the process whereby government has assumed responsibility for private losses as "Socialism for the rich and laissez-faire capitalism for everybody else." Socialization of risk means that the public is absorbing the costs of the bailout as the federal government bails out, backstops, and insures private losses.

In essence, critics of nearly all economic stripes have been critical of the U.S. domestic policy response. Social welfare Keynesians claim that the response has not accomplished the Keynesian long-term objective of stimulating demand, while neoconservatives feel that government policies are market distorting and that accumulated debt will eventually drain resources from private enterprise. The U.S. government's inability to act decisively to either nationalize banks (temporarily) or to let large financial institutions fail derived from the powerful role of market investors, including foreign investors, sovereign wealth funds, wealthy global capital investment vehicles (e.g., hedge funds), and pension funds. The U.S. policy response to act on behalf of the market (through credit infusions) without directly influencing market operations illustrates the subordination of the state to the principle of market autonomy.

This policy response will have significant effects on the state's government of civil society. Bailout and stimuli costs will adversely affect the fiscal health of all levels of government agencies. The projected deficit for fiscal year 2009 alone is already projected to reach $1.6 trillion ("Recession, Bailout"). The economist Michel Chossudovsky argues that the combined 2010 costs of the bank bailout and military spending constitute $2.3 trillion (in "America's Fiscal"). He observes these expenditures will consume almost the entirety of the U.S. Treasury's revenues for 2010 of $2.38 trillion. Continued defaults on commercial real estate, home mortgages, student loans, auto loans, and credit cards are expected, and these defaults will continue to place pressure on weak or zombified financial institutions, leading to further government expenditures to backstop financial losses in 2010 and 2011. Faced with the alternatives of slashing military spending or slashing social spending, it is very likely that the U.S. will elect to slash social spending to maintain its military-driven global leverage. Drops in federal funding for social welfare projects will shift burdens to more local levels of government, which are already reeling from the contraction of employment and consumption.

The U.S. government will, in the short term, finance the gap in real expenditures and revenue by selling debt, primarily in the form of treasury bonds. On February 19, 2009, Roubini warned of the risk of global sovereign defaults as a consequence of the accumulation of unsustainable debt by governments (Roubini "Worst"). Roubini did not specify the nations at risk for sovereign default. Usually, economists claim that the United States cannot default on its debt because, as the holder of the world's currency reserve, the United States can simply issue more currency to make good on its obligations. The inflationary tendencies resulting from monetizing debt make it easier to repay debt, but the value of the debt repaid declines, which rightly upsets creditors.

Despite its privileged status as world's currency reserve nation, the United States relies on foreign investors to absorb its debt. China, which holds the most U.S. debt to date, will be unable to finance the U.S. debt through consumption of Treasuries and other bonds because of its domestic emergency of mass unemployment wrought by sharply declining exports. China's consumption of Treasuries has already slowed (Bradsher), and in March 2009, the Chinese premier Wen Jiabao expressed concern about the $1 trillion of Chinese investments in U.S. debt given rising U.S. fiscal deficits (Wines). Other countries are also worried about risks to their U.S. debt holdings and are now publicly calling into question U.S. hegemony and the global, neoliberal order. In June 2009 China and Russia called for a more stable global economic system not dominated by a single power (Levy). Oil-exporting nations have also recently demonstrated less capacity and willingness to consume U.S. debt because of declining oil revenues. The strength of the U.S. dollar could be significantly undermined by resulting inflationary tendencies or by an unwillingness of foreign investors to continue financing U.S. debt.

The possibility exists that exploding U.S. debt will fundamentally change the neoliberal economic order that that has prevailed since the 1980s. While a new, more multicentric world order promises opportunities for emerging economies in the long term, it could have significant short- and midterm adverse effects both on poor populations in developing nations and on the U.S. populace generally. What follows briefly examines possibilities for a reallocation of global power before turning to address the immediate and potential long-term impact of evolving economic events on the U.S. populace.

A New Global Order?

The World Bank anticipates that the global economy will shrink in 2009 (Andrews A12). The IMF expects the global economy to stabilize in 2010, but it holds negative prospects for U.S. growth that year ("IMF: World Economy"). The IMF offers two scenarios for 2010: a "benign scenario" and a "downside scenario" (32). The benign scenario entails a "successful rebalancing of the global economy" fueled by strong consumption in east Asia, among other factors (32). The downside scenario entails lack of "fiscal consolidation" and high unemployment, leading to higher U.S. indebtedness and the potential risks of "disorderly unwinding" of the U.S. fiscal situation that will significantly impact developing nations reliant on export trade to the U.S. and nations possessing large dollar holdings/reserves (32).

The IMF's adverse scenario could result in a "sudden stop" of financial inflows to the United States, which would severely impact the U.S. economy and other nations in the world holding large quantities of dollar-denominated assets. A former IMF employee, Willem Buiter, describes the risk of "sudden stops" in nations such as the United States. and the United Kingdom:

> In a number of systemically important countries, notably the US and the UK, there is a material risk of a "sudden stop"—an emerging-market style interruption of capital inflows to both the public and private sectors—prompted by financial market concerns about the sustainability of the fiscal-financial-monetary programmes proposed and implemented by the fiscal and monetary authorities in these countries. For both countries there is a material risk that the mind-boggling general government deficits (14% of GDP or over for the US and 12% of GDP or over for the UK for the coming year) will either have to be monetised permanently, implying high inflation as soon as the real economy recovers, the output gap closes and the extraordinary fear-induced liquidity preference of the past year subsides, or lead to sovereign default.
>
> Pointing to a non-negligible risk of sovereign default in the US and the UK does not, I fear, qualify me as a madman. The last time things got serious, during the Great Depression of the 1930s, both the US and the UK defaulted *de facto*, and possibly even *de jure*, on their sovereign debt.

A "sudden stop" of external investment in U.S. debt would not only have a devastating impact on the U.S. economy but would also adversely impact all developing nations dependent upon trade with the United States or holding large quantities of dollars, which would lose value dramatically in the event of a sudden stop.

Kevin Phillips remarked in April 2009 that the 50-trillion-dollar meltdown of financial markets may ultimately reverse U.S. status as a superpower:

> Today's disaster stage of American financialization—the bursting of the huge 25-year, almost $50 trillion debt bubble that helped underwrite the hijacking of the U.S. economy by a rabid financial sector—won't be nearly so kind. It is already ushering in the reverse: a global realignment in which the United States loses the global economic leadership won in World War Two. The ignominy deserved by Wall Street after 1929–1933 is peanuts compared with the opprobrium the U.S. financial sector and its political and regulatory allies deserve this time.

The possibility exists that the U.S. economy will resist reinflation as job losses increase credit defaults and as corporations struggle to pay back loans in the face of declining sales.

In April 2009 the *Christian Science Monitor* ran an article titled "America: A Superpower No More: Decline Is Occurring More Rapidly Than We Think. It's Time to Embrace a New Agenda" (Rogers). The article compares the U.S. reliance on military interventions to open markets and foster "democracy" with China's (mercantile) economic approach to governing global relations:

> The price tag for the wars in Iraq and Afghanistan is in the trillions. Sun Tzu, the ancient Chinese military commentator, prophetically observed 2,500

years ago, "[W]hen the army marches abroad, the treasury will be emptied at home."

It remains a lingering American myth that US troops and warships can go anywhere and pay any price. Not so. The modern Chinese have discovered a better way. The *Washington Post* reports that the Chinese went on a shopping spree recently, taking advantage of fire-sale prices to lock up global supplies of oil, minerals, and other strategic resources for their economy. That amounts to a major economic conquest—without using a single soldier. By contrast, American efforts to secure oil have looked clumsy.

The author laments the loss of American "moral authority," but frames Middle Eastern and Chinese disenchantment with America as stemming from these countries' rejection of the "[W]estern democratic model." Disenchantment with democracy substitutes in the article for disenchantment with Chicago-style neoliberal principles of economic government.

The fiscal crisis in the United States and United Kingdom has emboldened other nations who dislike the U.S./U.K. neoliberal global order. In June 2009 the Russian president Dmitry Medvedev called for the use of national currencies in trade. This call implicitly constitutes a rejection of the U.S. dollar as the world's currency reserve ("Medvedev Calls"). During the June 16 summits in the Russian city of Yekaterinburg, the Shanghai Cooperation Organisation[1] and the BRIC nations (Brazil, Russia, India and China) promoted establishment of a "fairer world order" characterized by a "new global financial security system" and a "new super-national currency to handle mutual transactions" ("SCO, BRIC Urge Fairer World Order"). These calls represent an explicit disavowal of the global economic system that has significantly benefited wealthy populations within the United States and enclave elites abroad, even while that system has further impoverished much of the developing world and led to the gross exploitation of limited resources.[2]

Criticism of the neoliberal economic order has also emerged from relatively mainstream U.S. economists. In February 2008 Nouriel Roubini published an article in his online *RGE Monitor* titled "Worst Economic and Financial Crisis since the Great Depression Reveals the Weaknesses of the Laissez-Faire Anglo-Saxon Model of Capitalism." Roubini's article outlines the deficiencies in the extant neoliberal order, especially the "efficient market hypothesis" and the capacity for markets to self-regulate, and suggests directions for reforms he sees as absolutely necessary. Mark Thoma, also an economist, published a similar article titled "Goodbye, Homo Economicus" in the March 26, 2009, *RGE Monitor*. As these articles illustrate, even previously mainstream economists are questioning the viability of the neoliberal economic order that has prevailed for the last thirty years because it has produced vast global imbalances and allowed the United States to indulge in risky finance and excessive debt. What is particularly interesting about these criticisms is their willingness to engage with the fundamental tenets of liberal economic thought, including laissez-faire and homo oeconomicus.

Joseph Stiglitz explains global disenchantment with neoliberal eco-nomic policies developed and forcibly implemented by the United States and supranational agencies such as the World Bank and the IMF in terms of a fundamental failure to deliver basic improvements in standards of living for most of the world's populations. Rather than promoting the collective good, neoliberal policies contributed to growing inequality while causing more frequent financial crises. Often, "neoliberal" reforms simply masked outright economic exploitation, as Stiglitz explains in "Wall Street's Toxic Message":

> Free-market ideology turned out to be an excuse for new forms of exploi-tation. "Privatization" meant that foreigners could buy mines and oil fields in developing countries at low prices. It meant they could reap large prof-its from monopolies and quasi-monopolies, such as in telecommunications. "Liberalization" meant that they could get high returns on their loans—and when loans went bad, the I.M.F. forced the socialization of the losses, mean-ing that the screws were put on entire populations to pay the banks back. It meant, too, that foreign firms could wipe out nascent industries, suppressing the development of entrepreneurial talent. While capital flowed freely, labor did not—except in the case of the most talented individuals, who found good jobs in a global marketplace.

Stiglitz particularly emphasizes how neoliberal reforms forced developing nations to cut back their Keynesian social welfare expenditures in order to control deficit spending. These cuts harmed already-impoverished popula-tions but were held by neoliberal policymakers to be necessary for long-term economic stability. Stiglitz argues that developing nations therefore regard unprecedented U.S. and U.K. deficit spending as fundamentally hypocriti-cal. The United States and the United Kingdom promote neoliberalism until it becomes domestically inconvenient. Nations unhappy with the global neo-liberal order dominated by U.S. financial and military might are likely to support efforts by the BRIC nations to develop a new framework of global governance.

The future remains open, and it is impossible at this time to predict its contours. It is unclear what new economic logics and conditions will emerge, but certain trends are clear. The guiding logics of liberalism have come under assault for failing to deliver collective goods. At a practical level, the U.S. dol-lar will probably not be able to maintain its long-term status as the world's currency reserve. Emerging economies may no longer passively or willingly accept neoliberal development policies and programs dictated by the United States or U.S.-driven global governance institutions. Yet mercantile policies pursued by China and Russia are unlikely to replace neoliberal ones as long as global governance institutions such as the IMF and the World Bank pre-vail. Disorder could result from political contestations over the form and scope of governing logics. Disorder will grow as the resource depletions fueled by neoliberal logics amplify the desperation of entire populations.

Although disorder may offer opportunities for new actors, it will have significant adverse effects on the health and vitality of entire populations.

Disorder could dislocate the hegemony of global exchange media—the
dollar—and thereby dislocate the world's strategically devised imbalances.
The dollar's status enabled the United States to accrue unprecedented levels
of debt, contributing to global financial and trade imbalances. If the United
States loses its status as the nation possessing the world's currency reserve, it
will be forced to pay higher interest on its debt, limiting the resources available for investing in its population. It will struggle to pay its debt until it
replaces its consumption-driven and war economies with a productive economy that can compete effectively in a multicentered, globalizing world. Yet
this type of substitution requires centralized planning, a practice that runs
counter to the U.S. neoliberal orthodoxy. Consequently, the U.S. populace
may experience the implosion of "civil society" as the state sheds and priva-
tizes more and more of its liberal (social welfare) governmental operations.

UNEMPLOYMENT AND STATES' FISCAL CRISES

The U.S. economy is primarily driven by consumption: up to 70 percent of
the recent GDP has been consumption related ("Facts on Policy"). Americans
who have lost home equity, stock market wealth, and access to easy credit
will not be able to maintain previous consumption levels. Americans lost
approximately 18 percent of their wealth in 2008 because of plunging stocks
and home values (Kalita A8). They are expected to lose still more in 2009.
The significant implosion in consumption that is occurring now will not
fully abate until the growing unemployment rate and reduced work hours
reverse. These reversals are not likely to occur until more economic activity shifts away from discretionary services, construction, and retail, since
these economic sectors are suffering from significantly dampened consumer
demand. The U.S. must develop new productive sectors capable of creating
jobs to replace the jobs lost in the imploding and job-hemorrhaging housing
and consumer-related sectors of the economy.

In the last quarter of 2008, the U.S. GDP shrank at an annualized rate
of 6.2 percent (Shin and Irwin A1). The first quarter of 2009 witnessed
another 5 percent contraction. This degree of decline is not expected to
continue, but the jobs lost are not expected to reappear either. The official,
national unemployment figure at the end of June 2009 was 9.5 percent.
The unemployment figure excludes workers employed in part-time jobs but
seeking full-time jobs and unemployed individuals who lack unemployment
benefits. The broadest measure of unemployment, which still excludes dis-
couraged workers, exceeded 16 percent in June 2009 (U.S. Department of
Labor "Employment Situation").

Young workers and lower-income workers initially bore the brunt of the
job losses (Eckholm "Working Poor"). Graduating college students face
"low wages for years to come" (S. Murray A1, A11). Low-income citizens
employed in the service, manufacturing, and housing sectors who have been

significantly impacted by job losses face "fraying" social nets (Eckholm "Safety Net"). Older workers are finding it increasingly difficult to replace lost jobs at comparable incomes. Many older workers are returning to the workforce after losing much of their retirement savings. The U.S. economy lost 7.7 million jobs since the recession began in 2007 (Lahart A1). The U.S. economy must add 100,000 to 125,000 jobs a month simply to keep pace with population growth.

In June 2009 an article in *USA Today* described deteriorating work conditions for the employed. The article reports: "People who still have jobs are faring worse than at any time since the Great Depression, a *USA TODAY* analysis of employment data found. Furloughs, pay cuts and reduced hours are taking a toll on workers who so far have escaped job cuts" (Cauchon). Employed workers logged only an average of 33.1 hours a week. Part-time work as a percentage of employment exceeds all records, and overtime is at a record low, according to the Bureau of Labor Statistics (Cauchon). The article reports that unemployment for workers forty-five years and older is higher than any time since the Great Depression. Fewer citizens have health insurance as unemployment rises and as companies cut benefits (Alonso-Zaldivar; Petrecca).

Small businesses and large corporations alike tied to the consumption economy are struggling in a recession now projected to continue into 2010. Publicly traded companies will seek to offset losses by gains in productivity, which are likely to be achieved by laying off employees. Unemployment benefits are being challenged by employers as unemployment grows (Whoriskey A1). Remaining employees desperate to retain jobs will probably work more for less.

In this context of market struggle, businesses of all sizes have pressed for and received tax cuts. For instance, in the middle of the financial crisis, in November 2008, the U.S. Treasury issued new tax policy guidelines that reduced the tax burden on banks by up to $140 billion, thereby culminating eight years of federal tax cuts for the largest and most powerful U.S. corporations (Paley A1). While tax cuts may be appropriate for struggling small businesses, many large corporations entered the recession with huge capital assets accumulated from twenty years of efficiency gains achieved through offshoring and automation and through the growth of financial services. Indeed, U.S. corporations enjoyed perhaps the greatest profit cycle in the post–World War II period until the beginning of its decline in late 2006/early 2007 (R. Nutting). In 2005 the *Economist* observed, "Last year, America's after-tax profits rose to their highest as a proportion of GDP for 75 years" at "the expense of workers" ("Breaking Records").

Declining corporate and business profitability and the capacity to write off losses will further reduce tax collections by states and by the federal government. Declining personal income will also eat into tax revenue. A June 18, 2009, *Wall Street Journal* article reports that state income tax revenue fell 26 percent in the first four months of 2009 compared with the same period in 2008 (Alini A4). States such as Arizona, California, Florida, and Nevada already face severe fiscal challenges in the struggle to balance budgets for the fiscal year 2010. The Beacon Hill Institute projects state revenues to decline by an average of 6.3 percent over 2009; however, this projection is probably optimistic given the severity and acceleration of first-quarter revenue declines. In June 2009 states faced a cumulative shortfall of $230 billion through 2011 (Weisman A2).

States are having difficulty investing in their infrastructures and maintaining educational outlays and social services because of their severe financial situations. Federal stimulus funding is helping to offset some state cuts in education and social services, but stimulus money will dissipate in fiscal year 2011. Construction activity is up slightly with stimulus spending on projects such as federal highways, but this activity is also temporary. The U.S. federal government's investments in states are not nearly enough to offset declining private and local government expenditures. Consequently, economists predict a "W"-shaped or "L"-shaped "jobless" recession. Most recently, Robert Reich opined that no recovery is possible as long as current governing logics and practices prevail:

> My prediction, then? Not a V, not a U. But an X. This economy can't get back on track because the track we were on for years—featuring flat or declining median wages, mounting consumer debt, and widening insecurity, not to mention increasing carbon in the atmosphere—simply cannot be sustained.

A global look at the resource depletions and devastations wrought by neoliberal logics supports Reich's contention about the impossibility of "recovery." Yet efforts to develop new logics of economic and social government, as Reich suggests, face seemingly intractable inertia and resistance. The extant debt trap stemming from neoliberal financialization and consumption is like quicksand, consuming resources and biovitalities.

AMERICAN FAMILIES AND CHILDREN IN THE RECESSION

How will the U.S. recession shape American families and the government of childhood? Economists predict that Americans will enjoy less access to consumer goods because of lost income, equity, and international leverage. Middle-class and working-class parents will no doubt witness significant public disinvestments in public schools and communities (e.g., parks, community centers, libraries) as revenue-strapped states, counties, cities, and school districts confront declining tax revenues caused by the recession. The "great risk shift" described by Hacker will be amplified as companies shed benefits, such as 401(k) matching. The risk shift seems unlikely to be reversed by public sector investments given the fiscal pressures facing government as it confronts reduced global power, reduced revenues, rising domestic deficits, and an aging and increasingly unhealthy population. The liberal space of civil society will slowly implode, and this implosion will erode the nation's collective biovitality.

Many Americans will be impoverished by the loss of jobs and home equity. The Center for Budget and Policy Priorities reported in November 2008 that the recession could cause large increases in poverty and push millions more into deeper poverty (Parrott). The center predicted that a 9 percent unemployment rate would cause approximately 10 million more to enter poverty. In July 2009 the unemployment rate was 9.5 percent.

Hunger is rising among American households because of the recession (Koch 3A). A survey released by the group Feeding America at the close of 2008 found that 36 percent of low-income households claimed they ate less or skipped meals because they lacked money for food, and 40 percent say they chose between paying food and paying utilities over the past year. The study claims that 17 percent of American children under the age of five, totaling more than 3.5 million children, face hunger (Spencer). The data collected in this survey cover the period ranging from 2005 to 2007. The president of Feeding America summed up current conditions by stating: "We've never seen anything like this...We're seeing more people come (to food banks) who've never come before" (quoted in Koch 3A). The U.S. Department of Agriculture reports that the number of people receiving food stamps went from 26.9 million in September 2007 to a record 31.6 million in September 2008 (Koch 3A). No data are yet available for 2009, but the expectation is that hunger and food insecurity continue to rise. A 2009 report on child and youth well-being argues that parents are likely to substitute more healthful, but expensive, food choices with less nutritious, but also less expensive, options, exacerbating childhood obesity and adversely impacting children's overall health (Land 4)

Rising home foreclosures are producing crises for displaced families. For instance, New York City homeless shelters saw a 40 percent increase in homeless families in the period ranging from July to November 2008 compared with July to November 2007 (Sard). Homelessness is growing even among two-parent households (Jenkins A1). Homeless families with children are already being turned away from overcrowded shelters.

The risk shift in retirement enabled by 401(k) funds and rising equity in homes has ended disastrously for a large near-retirement baby-boom generation. Comfortable retirement may become impossible for the vast majority of Americans given the tremendous financial losses experienced by public and private pension funds. Aging boomers who have lost over half of their retirement investments will fight to preserve their Social Security and Medicare benefits. Mismanaged public pension programs crippled by investment losses will pose significant problems for public officials, who must choose between honoring promised retirement benefits and investing in education, health, and social welfare programs for children.[3] Aging citizens facing deteriorating welfare prospects will agitate strongly for retaining benefits. Organized under the national umbrella of the American Association of Retired Persons (AARP) and similar organizations, older populations will no doubt be more successful than children in mobilizing support for their interests. Additionally, older Americans faced with deteriorated investments are remaining in the workforce rather than retiring, whereas formerly retired workers are now attempting to reenter the workforce (Evans "Ranks of Older" A2). These trends heighten job competition.

America is already an aging and adult-geared society. The boomers' retirement is amplifying that trend. Lack of identification with, and sympathy for, America's minority children will no doubt play a role in retirees' unwillingness to reinvest in the populace. Already, retirement-only incorporated cities such as Sun City and Sun City West have been devised as strategies for avoiding paying school taxes.

America's youthful populations will be forced to assume more risks even if the society as a whole slowly discards some neoliberal risk frames. Health coverage and Social Security outlays for older populations will leave little federal moneys available for children, leaving that burden to fiscally damaged states whose budgets will take years to recover from the shocks of the recession. America's unwillingness to cut military spending and domestic policing will ensure that social welfare services for the poor are cut. In the absence of radical changes in government logics, policies, and strategies, U.S. children will no doubt experience declining economic opportunities.

Children in public schools face more crowded classrooms as states slash budgets in the face of rapidly declining sales revenues (Lav and McNichol). Education constitutes the largest budget item for most states and will therefore experience significant budget reductions as states encounter ongoing fiscal meltdowns. School districts will have little choice but to cut teachers, given the lack of discretionary spending in most districts' budgets for operations. Affluent populations will turn to private school and will therefore be unwilling to support tax increases aimed at helping struggling public schools. Social welfare for children seems to be doomed to death by a thousand cuts in the absence of significant changes in the organizing logics of American society.

Household stress and anxiety will no doubt increase as parents face pay cuts, unpaid furloughs, job losses, and lost savings and retirements. Parents faced with fiscal constraints may forgo divorce. While this decision may reduce economic stress, it can simultaneously produce more domestic stress as parents fight and displace anger upon children. These trends are likely to have adverse impacts on the health and vitality of the population. First Focus reports in a December 2008 report, *The Cost of Doing Nothing*, that the recession will cause an additional 2.6 to 3.3 million U.S. children to fall into poverty. The report financializes the social impact of this rise in poverty by observing that an additional 3 million children in poverty translates into at least a $1.7 trillion loss to the overall U.S. economy.

RADICAL POLITICS

The economic events of 2008 and 2009 have shocked and confused American citizens. The failures of AIG and Bear Stearns have been followed by a never-ending media stream of reports announcing bank failures, corporate bankruptcies, home foreclosures, and rising unemployment figures. Something is not "right" in America, but who or what is to blame? The *Christian Science Monitor* points to the loss of U.S. moral authority and the subsequent disenchantment with Western-style democracy. This narrative is but one of many narratives circulating throughout society that attempt to explain the seemingly imminent collapse of the United States as the world's only superpower. As anger mounts within the United States, the narratives created to justify transformed social and economic conditions may be less benign.

Americans are accustomed to consuming cheap goods that have been enabled by the dollar's status as the world's currency reserve. American's are used to paying relatively, globally speaking, low prices for the gas necessary for transportation in most American cities. Americans of all social classes are accustomed to American military might and broadly believe in the tenets of American manifest destiny. Most of all, Americans believe in a consumption-driven "good life." This good life is held to be broadly accessible, as outlined in the mythic narrative of American self-transformation. All of these lifestyle expectations have become increasingly tenuous given the financial meltdown of the U.S. economy. Moreover, many of these assumptions, specifically the availability of low-cost oil and the endless consumption of cheap goods, face finite limits. Scarcity looms within global resource horizons.

The meaning of events will be constructed, deconstructed, and reconstructed in politicians' rhetoric, print media editorials, talk radio, and Internet blogs. The future remains open, and it is impossible to disclose political trajectories. Perhaps Americans will protest growing unemployment and government bailouts for the nation's wealthiest individuals and corporate interests. Perhaps radical politics will emerge. Perhaps authoritarianism will grow also. Authoritarianism often arises to suppress the resistance produced by faults and contradictions in governing logics.

Policymakers and authorities in the advanced economies may seek to displace blame and risk upon vulnerable populations domestically and abroad. Wars often unify national populations by providing meaning in times of uncertainty until someone stops to count the dead and the costs of war for the living. Global populations who reject the neoliberal order or simply compete for scarce resources may be cast as enemies of democratic capitalism and targeted for military intervention. Already, impoverished populations emerge as "security risks" in U.S. defense rhetorics (Warrick A1). This framework, which circulates widely in the U.S. electronic and print news media, will discourage any efforts to reduce U.S. deficits by cutting military spending. Repressive apparatuses purporting to protect the national race will thrive in this type of environment, displacing more pastoral, democratic, and egalitarian biopolitical apparatuses.

GLOBAL CHILDREN

Economic events may very well amplify "security risks" and further upset international support for a neoliberal economic order that requires poorer nations to enable the consumption of wealthy nations. Economic contraction is already exacerbating global poverty and may amplify regional conflicts over scarce resources. The World Bank reported in February that up to 53 million more people would become mired in poverty because of soaring food and fuel prices catalyzed by the financial crisis ("Economic Crisis Set to Drive"). This increase would bring the total number of people living on less than $2 a day to over 1.5 billion. Children will be disproportionately impacted by this crisis, as the United Nations news center reports:

> The global financial crisis sweeping through Wall Street and the European banking sector will touch the lives of the world's most vulnerable, pushing millions into deeper poverty and leading to the deaths of thousands of children, according to a new United Nations study.
> Reduced growth in 2009 will cost the 390 million people in sub-Saharan Africa living in extreme poverty around $18 billion, or $46 per person, warned the report by the UN Educational, Scientific and Cultural Organization … New estimates for 2009 suggest that lower economic growth rates will trap 46 million more people on less than $1.25 a day than was expected prior to the crisis, for a total of an extra 53 million trapped on less than $2 a day, on top of the 1.37 billion before the current crises. ("Financial Crisis to Deepen")

The World Bank observed that approximately 40 percent of the 107 developing nations were highly exposed to the effects of the crisis ("Financial Crisis to Deepen").

The outflow of currency from developing nations to affluent ones, coupled with the decline in global consumption of exports, threatens to imperil developing economies that focus on export-oriented production (Goodman A1, A15). A World Bank report issued in June 2009 notes further deterioration of the global economy and expresses concern for the "damage" done to low-income countries by the decline in export and the outflow of capital, which are together rendering it difficult, if not impossible, for many developing countries to finance needed imports such as basic commodities (World Bank "Global Development" 6–9). Increased global risk aversion has made it difficult for companies and governments in developing economies to raise capital (10). Worse, depreciation of currencies in the developing world has increased the "local currency cost of servicing dollar-denominated debt" (World Bank "Global Development" 18). This means that developing nations will have to pay more of their own currency to service foreign debt denominated in dollars unless debt forgiveness or refinancing occurs.

The only bright trend observed by the World Bank is a drop in the prices of basic food commodities to December 2007 levels (16). Global commodity speculation in 2008 had contributed to dramatic increases in food prices and to hoarding, which together caused food prices to rise significantly (A. Davis A4; Hookway and Etter A7). Food riots resulted in many developing nations as rising food prices produced desperation. Although the World Bank believes food prices will remain stable (World Bank "Global

Development" 17), a number of financial forecasters predicting food shortages are encouraging traders to invest in agricultural products, as illustrated by this investment article's title, "The Real Crisis is Food: Beginning of the Bull for Agriculture" (Summers). Speculation by affluent investors will drive up prices and hoarding. The poor will starve.

Increased impoverishment and nutritional deficiencies could easily stem from the global economic recession in the absence of deliberate efforts by nation-states and global governance institutions to pursue policies and programs beneficial for developing nations. Faced by severe fiscal shortfalls, (formerly) wealthy nations are anticipated to make drastic cuts to already low levels of foreign aid (see Macfarquhar). Nongovernmental organizations (NGOs) have been severely impacted in this downturn and will have fewer resources available for their various development projects. Additionally, remittances to poor nations from the United States and the United Kingdom are down dramatically as construction and service-sector jobs evaporate in developed economies. Developing nations facing default on their loans or fiscal crisis may be forced to turn to global governance institutions such as the World Bank and the IMF despite deep skepticism and disillusionment with their methods and motives.

These developments are likely to amplify growing disillusionment with the neoliberal global order in emerging economies. Local populations further impoverished by collapsing exports, lack of access to equity, and government austerity may very well demand that their nations resist a global order that is both unraveling and demonstrably counterproductive for developing nations' economic interests. Civil unrest may ensue. Developing nations that retreat from the neoliberal economic order by violating trade covenants, by nationalizing industries that had been forcibly privatized in the past, by moving away from the dollar as their currency reserve, by defaulting on debt payments, and so on, face covert or overt military reprisals. The neoliberal global order that has prevailed for the last three decades is unlikely to withstand collective resistance from impoverished populations without neoliberal authorities resorting to brute force.

POLICING POPULATIONS THROUGH "STRATEGIC SHOCKS"

Growing unemployment and impoverishment across the globe are likely to produce new "security risks" for the prevailing, but deeply damaged, global economic order. These risks are exacerbating the risks already anticipated to occur because of resource and food scarcities produced by global climate change (Pincus A2). In November 2008 the *Strategic Studies Institute* issued a report titled "Known Unknowns: Uncoventional 'Strategic Shocks' in Defense Strategy Development," authored by Nathan Freier. The report summary explains its objectives to anticipate and develop contingency plans for unconventional "dangerous future shocks" that "manifest themselves in ways far outside established defense convention" (vii). Although most of the shocks are anticipated to be "nonmilitary in origin and character," Department of Defense (DoD) planning is recommended. The types of shocks included in this planning document include the following:

> Threats of context might include but are not limited to contagious un- and under-governance; civil violence; the swift catastrophic onset of consequential natural, environmental, and/or human disaster; a rapidly expanding and uncontrollable transregional epidemic; and the sudden crippling instability or collapse of a large and important state. Indeed, pushing at the boundaries of current convention, it would be prudent to add catastrophic dislocation inside the United States or homegrown domestic civil disorder and/or violence to this category as well. (17)

The report explains that most of these "contextual threats" are the origins of shocks, since they operate as triggers or catalysts. The DoD will be forced to "fundamentally reorient strategy, capabilities, investments, and concepts in response" (18). Shocks emerging within or outside the United States would force the DoD "to radically re-role for domestic security, population control, consequence management, and stabilization" (18).

Paths to domestic civil violence that would require the DoD to reorient priorities "to defend basic domestic order and human security" include deployment of "weapons of mass destruction, unforeseen economic collapse, loss of functioning political and economic order, purposeful domestic resistance or insurgency," and so on (32). Civil violence might require the "use of military force against hostile groups inside the United States" (33). Moreover, the DoD "would be, by necessity, an essential enabling hub for the continuity of political authority in a multi-state or nationwide civil conflict or disturbance" (33). In other words, the report outlines when and how the DoD would assume responsibility for direct domestic governance.

The report considers the possibility that hostile states or nonstate actors might "combine hybrid methods effectively to resist U.S. influence in a non-military manner" and notes that "this is clearly an emerging trend" (33). The report cites the China-Russia axis as capable of holding "American interests at risk" (33). The report describes how nonmilitary moves by political-economic adversaries could limit "American freedom of action" (34). Under the scenarios envisioned in the report, U.S. military forces would "hold little promise for reversing adverse political and economic conditions" (34).

When one reads this report carefully, it becomes clear that its primary objective is to describe risks resulting from the increasing vulnerability of the neoliberal economic order that has prioritized, and been predicated upon, U.S. economic and political interests. The report implicitly warns of imminent economic resistance by BRIC and SCO (Shanghai Cooperation Organisation) nations. The report formulates their economic resistance as a fundamental security threat requiring DoD to articulate new problem solution frames for ensuring maintenance of what is euphemistically described as U.S. "freedom of action" (34). The report fails to describe how the DoD might respond in the international context to anticipated threats but concludes simply by urging "preemptive examination of the most plausible 'known unknowns'" for "strategic preparedness" (36). As outlined above, the report advises the unconstitutional use of military forces within the United States to ensure "domestic tranquility" (33).

Populist resistance in the United States in response to growing unemployment and impoverishment is therefore very likely to elicit state repression. The American Civil Liberties Union (ACLU) reports that DoD terrorism training materials currently employed describe public protests as "low-level terrorism" ("ACLU Challenges"). Additionally, the Pentagon plans to have 20,000 uniformed trained troops inside the United States by 2011, purportedly to help state and local officials respond to a terrorist attack or some other domestic catastrophe (Hsu and Tyson A1). The *Washington Post* reports resistance to this plan:

> Domestic emergency deployment may be "just the first example of a series of expansions in presidential and military authority," or even an increase in domestic surveillance, said Anna Christensen of the ACLU's National Security Project. And Cato Vice President Gene Healy warned of "a creeping militarization" of homeland security. (Hsu and Tyson A1)

The possibility exists that authoritarian totalitarianism will emerge if economic conditions continue to deteriorate significantly within the U.S.

Anticipating how the DoD and other government agencies will respond abroad to "strategic shocks" produced by continuing economic upheaval is difficult, because most responses are likely to operate covertly without media coverage. Some commentators argue that the U.S. Treasury and stock markets are already being manipulated to protect against collapse caused by declining demand and covert hostile market actions by foreign nations or actors. Indeed, the Pentagon purportedly held economic war games modeled upon real-world conditions in the spring of 2009 (Egan). The United States apparently will not willingly or gracefully accept any significant decline in its status as the global superpower.

CONCLUDING THOUGHTS

This book has described the biopolitical risks posed to and by children across the twentieth and into the twenty-first centuries as conceptualized in changing logics of societal government. Poor populations within the United States have nearly always been regarded as potentially dangerous, but the problem-solution frames used to represent and govern poor populations have changed across time. Children and childhood have stood at the nexus of social policies and attitudes toward poor populations for the last 150 years.

Children have served in the Western, modern imagination as a concrete index of future conditions. For the last 150 years, Western authorities have defined and measured childhood characteristics and dispositions in order to envision and control future social conditions. This project has, of course, been in many ways deluded because of the (inevitably) situated and interested nature of measurement devices and because of the ultimately ungovernable effects of human performances. Yet twentieth-century efforts to govern childhood's present in order to govern society's future did encourage social welfare practices that enhanced the biovitality of the population.

Foucault's works, especially *The History of Sexuality* and *Psychiatric Power*, examine the power effects of even benevolent-seeming biopolitical practices. Foucault's analysis of government in "Governmentality" points to the circulation of security mechanisms that elicit and channel the population's life forces, producing new micropolitics that act upon and through human performances. There can be no doubt that biopolitics infuses all pedagogies and childhood-governing practices. Yet this indisputable fact does not mean that all forms of power are equal in terms of their effects. Although infused with power relations, social welfare practices created spaces for resistance as evidenced by the formulation of child rights, the civil rights movement targeting minority and women's rights, the environmental movement, and the disability rights movement. These movements all drew upon the child—ranging from anonymous disabled children to iconic children such as Ruby Bridges—to build popular support for more egalitarian logics and strategies of government. While images of children can be used for many purposes, as evident in chapter 5, concern for the problem-solution space of childhood has been central to social welfare rhetorics and apparatuses for the last century. The child's constitution as a subject of rights within social welfare logics and rhetorics promoted security apparatuses that hedged against child starvation, malnutrition, and gross abuse and mistreatment.

Over the last thirty years, the ascendancy of neoliberal and conservative logics of government has produced a "great risk shift" that promises to unravel much of what remains of the collectivized, "social welfare" risk-management apparatuses in the United States. Neoliberal and (neo)conservative problem-solution frames represent childhood dependency as threatening society by producing market exclusions. Dependent children neither participate fully in the marketplace of goods nor do they (usually) grow up to achieve the idealized adult entrepreneurial financial status fetishized in neoliberal and (neo)conservative market fantasies. Over the past thirty years, policymakers in the United States have regarded a "culture of poverty" as responsible for producing dependency, for impinging against enterprise and competition, and for reproducing poverty. This environmental approach encouraged policymakers to emphasize parenting and early education as strategies for intervening in the reproduction of poverty. However, these strategies failed to address the economic conditions of possibility for poverty. Moreover, these strategies were partial in application, and educational funding for poor populations failed to compensate for dramatic inequalities in educational opportunities.

The purported "failure" of social welfare interventions measured against absolute standards of success (i.e., the elimination of dependency) therefore contributed to the growing influence of hereditary accounts of intelligence and aptitude. Hereditary accounts gained credence in the context of the nation's collective enthrallment with a simplified gospel of genomic science, which circulated widely in popular media. Policymakers stressed the failure of social welfare logics and hinted at the heritability and enculturation of dependency when implementing neoliberal/neoconservative reforms, which shifted risk to individuals.

Neoliberal logics and strategies of government have polarized populations within the United States and across the globe by increasing the availability of resources and opportunities for small population segments while impoverishing and marginalizing population majorities. The global financial crisis discussed in this chapter stemmed directly from neoliberal market excesses and inequities, which were often backed by authoritarian repressive apparatuses such as domestic policing, international arms transfers, and military occupations. This chapter explored trends in the material effects produced by this crisis in the context of neoliberal logics and policies that have shifted risk to local governing units (e.g., states, counties) and individuals. Impoverishment grows within the United States., contributing to the crisis, which further destabilizes export-oriented economies in the rest of the world.

Crises produce opportunities by dislocating prevailing logics. Perhaps this crisis will allow the emergence of more collectivized governing logics within the U.S. that do not marginalize and punish those forced to bear the burdens of logics that individualize risk. Perhaps childhood will emerge again as a category of need demanding social welfare government logics capable of eclipsing and overriding logics of risk that reward exploitation of natural and human resources. Perhaps the subject of rights will be prioritized over homo oeconomicus.

In *The Birth of Biopolitics,* Foucault observes that liberalism produced two distinct subject positions that impinged against centralized state power, the subject of rights and homo oeconomicus. The political subject of rights and the economic subject of the market were never reducible, but they often operated commensurably within liberal rhetorics and practices of government (292). At times of crisis, the fundamental tensions between and across these two formulations often appear, as evidenced by the conflict between labor rights and employer rights during the Great Depression. Neoliberalism sought to resolve potential tensions by collapsing the subject of rights into the enterprising subject of the market through the extension of market logics into all realms of societal existence (see 268–69). This reduction contributed to the discourse and practices of financialization that shaped middle-class childhood as discussed in chapter 3. Yet the unresolved tensions in this reduction were evident by the global childhood flows of immigrant children caused by economic circumstances and conflicts. In the U.S. and abroad, these immigrant children and their parents are denied legal rights of personhood, although employers often welcome their economic labor, as explained in chapter 5. The eclipse of the subject of rights by the subject of the market enables detention, incarceration, and deportation of individuals who have no legal standing of personhood, only economic energies available for exploitation.

Paradoxically, the ascendancy of the subject of the market—originally intended as a limit on excessive governmentality (320)—has had the effect of increasing the scope and circulation of repressive apparatuses throughout society. The over 800 U.S. military posts abroad illustrate how force has been used to promote and solidify the purportedly free subjects and operations of the market. Neoliberalism has ultimately destroyed precisely that which it sought to promote and protect: liberal freedoms defined both in terms of market autonomy and personal agency. Market and state authoritarianism increasingly supplement liberal freedoms and securities. Childhood populations posing threats to fragile market and social configurations shaped by neoliberalism may be increasingly targeted by authoritarian apparatuses, as the crisis and ongoing resource depletions produce vast social dislocations.

This project concludes with the observation that liberalism is in throes of an unprecedented crisis. Liberalism's internal contradictions and tensions have produced a financial crisis that points to, and derives from, the fundamental unsustainability of a governing logic that ultimately presumes the invisibility of the collective good while privileging individual pursuit of resources. Children, who have stood at the nexus of liberal governmental operations, face new conditions and experiences shaped by this immediate financial crisis and the resource scarcities exacerbated by the logic of market autonomy. The future remains open, but one conclusion is certain: childhood formulations that emphasize dependency/vulnerability while demanding social welfare accountability will be challenged by new economic and environmental problem-solution frames that emphasize scarcity and that individualize risk management.

NOTES

I   INTRODUCTION TO BIOPOLITICS, RISK, AND CHILDHOOD

1. Notable educational and sociological scholars whose work extends Foucauldian analyses to the study of children and childhood include, among others, the following: Marianne N. Bloch, Kerstin Holmlund, Ingeborg Mocqvist, and Thomas Popkewitz (1–334); Marianne N. Bloch, Devorah Kennedy, Theodora Lightfoot, and Dar Weyenberg (1–229); Roger Deacon (84–102, 435–458, 1–306); Henry A. Giroux (587–620); Kenneth Hultqvist and Gunilla Dahlberg (1–292); James Marshall (1–221): Mark Olssen (1–216); Michael Peters (1–263), Thomas Popkewitz (1–138); Roxanna Transit (1–198).

2. In Chapter 11 of the *Birth of Biopolitics*, Foucault differentiates the subject of the market, homo oeconomicus, from the juridical subject of juridical will. The latter is produced by the state's apparatuses and political/philosophical reflections on rights. Foucault argues that these two subjects are often commensurable, but are not reducible (273–76). Unlike the juridical subject, the economic subject of interest, homo oeconomicus, is "never called upon to relinquish his interest" (275), in large part because of the assumption that his egoistic pursuit of interest benefits others, as described by Adam Smith in terms of the beneficence of the market, the "invisible hand" (278). Foucault deconstructs the premises of the invisible hand, noting that Smith's formulation requires that all market actors be blind to the "collective good" (279). Hence, "invisibility" of the collective good is critical to Smith's theory; also important to Smith's view is the premise that government must interfere with the pursuit of personal self-interest or it risks inadvertently destabilizing the collective good (280). These principles, which assume a metaphysical market hand, explain liberalism's fundamental aversion to planning.

3. Keynes's *The General Theory of Employment, Interest, and Money* (1965/1936) legitimized state intervention in the field of economy, offering a critique of laissez-faire economics that punctured the ideology of market autoregulation. In order to increase economic demand, Keynes proposed that some external force—such as that constituted by state action—be brought to bear upon factors shaping demand's contours, including unemployment and education. Keynesian economics emphasized the state's role in ensuring individuals' "positive freedom" by collectivizing societal risk.

2   A Genealogy of Family Life and Childhood Governance

1. Nikolas Rose's *Powers of Freedom* (1–284), Jurgen Habermas's *The Structural Transformation of the Public Sphere* (1–250), and Mitchell Dean's *The Constitution of Poverty: Toward a Genealogy of Liberal Governance* (1–219) all provide excellent accounts of freedom and the relationship between the public and private spheres.
2. Louis Althusser described interpellation as the process of transforming individuals into social subjects. "Social subjects" are social roles constituted by structures such as class, race, and gender. Althusser uses the metaphor of "hailing" to describe this process of interpellation (162).
3. The sanitarians were early and mid-nineteenth-century urban sanitary reformers in England and the United States who sought to prevent the circulation of diseases linked to poor living conditions and poverty.

3   Risk, Biopolitics, and Bioeconomics

1. Rima Apple's *Perfect Motherhood: Science and Childrearing in America*, Julia Grant's *Raising Baby by the Book: The Education of American Mothers*, Ann Hulbert's *Raising America: Experts, Parents, and a Century of Advice about Children*, and Peter Stearns's *Anxious Parents: A History of Modern Childrearing in America* provide comprehensive and nuanced accounts of U.S. parenting advice.
2. Although Gallup reports that the public expresses uncertain and ambivalent feelings about private school vouchers when polled (http://www.gallup.com/poll/1612/Education.aspx).
3. Foucault describes technologies of the self as practices used by the self, upon the self (or upon relational others). Technologies of the self are biopolitical practices (Foucault *Technologies* 16–49).
4. A recent boycott of public schools in Chicago illustrates the discrepancy between public school funding in inner-city areas compared funding in outer suburbs. Chicago spends $10,000 per student annually compared with more than $15,000 per student in the northern suburb Winnetka. Likewise, Kozol argues in the *Harper* article, published in 2005, that "the present per-pupil spending level in the New York City schools is $11,700, which may be compared with a per-pupil spending level in excess of $22,000 in the well-to-do suburban district of Manhasset, Long Island."

4   Biopolitical Sorting: Comparing Neoliberal and Social Welfare Problem-Solution Frames

1. However, lead was not the only contaminant to raise alarm in this period. For example, a 1966 study on increased rates of leukemia in children cited benzene, tobacco smoke, insecticides, paint, and hairspray among other possible threats to health (Hernandez and Tokuhata 604).
2. This author surveyed two indices of citations to examine research on and about children, hyperactivity, and delinquency throughout the twentieth century. Those indices were JSTOR and PsycINFO.

3. In technical language, heritability is "an estimate of the contribution of genes and the environment to *individual differences* in any given trait" (Marcus 8). Heritability doesn't specify genetic influence to a trait (e.g., intelligence) but rather measures "what percentage of the variation in that trait can be attributed to" genetic influence (9).

4. Initially heading the program was Frederick Goodwin, whose careless public remarks comparing inner-city youths with violent oversexed monkeys in the wild eventually doomed both his career and the Violence Initiative (R. Wright 257–65). Although Goodwin's remarks provoked outrage, they also further cemented public perception that some individuals were inherently more "risky" or dangerous than others.

5 Biopower, Security, and Development

1. Michael Hogan's *Cross of Iron: Harry S. Truman and the Origins of the National Security State 1945–1954* (1–482) explains these developments in greater detail.

2. Domestic anxiety about threatening foreign agents facilitated what Kiracofe describes as the "imperial Presidency." The classic text on the subject is Arthur Schlesinger's *The Imperial Presidency* (1–500). Under the administration of G. W. Bush, the United States probably came closest to abandoning those liberties defined as central to "democratic" societies, including habeas corpus and free speech (see Cohen and Wells 1–240). The rise of U.S. state sovereignty, embodied in the form of an imperial presidency or administration, and the growth of the state's police apparatuses point to the reassertion of repressive power, as detailed by F. William Engdahl (1–245) and Chalmers Johnson (1–367). Authoritarian state power justified in relation to security threats points to the fusion of biopolitics and sovereignty in the modern era. This authoritarianism shapes American childhood and the childhoods of children abroad (e.g., see Henry A. Giroux "The New Authoritarianism").

3. See Ronald Greene's *Malthusian Worlds: U.S. Leadership and the Governing of the Population Crisis* (1–250), Wendy Kline's *Building a Better Race: Gender, Sexuality, and Eugenics from the Turn of the Century to the Baby Boom* (1–164), and Alexandra Minna Stern's *Eugenic Nation: Faults and Frontiers of Better Breeding in Modern America* (1–216).

4. For a discussion of the constitution of American manhood in the Cold War nexus, see K. A. Cuordileone's *Manhood and the American Political Culture in the Cold War* (1–246).

5. For a recent discussion of the U.S. political economy of militarism, see Ismael Hossein-Zadeh's *The Political Economy of U.S. Militarism* (1–258) and Nick Turse's *The Complex: How the Military Invades Our Everyday Lives* (1–272).

6. For excellent discussions of the privatization of aid, see Michael Edwards's *Just Another Emperor: The Myths and Realities of Philanthrocapitalism* (1–92), Arturo Escobar's *Encountering Development: The Making and Unmaking of the Third World* (1–226), and Julia Elyachar's *Markets of Dispossession: NGOs, Economic Development, and the State in Cairo* (1–279).

7. David Lyon's *Surveillance Society: Monitoring Everyday Life* (1–154) and Torin Monahan's edited *Surveillance and Security: Technological Politics and Power in Everyday Life* (1–292) provide comprehensive introductions to the practices and technologies of the U.S. security state. Michael Peters's edited *Education,*

Globalization, and the State in an Age of Terrorism describes the implications of these surveillance networks and technologies for the education of children in the context of the war on terror (1–258), while Joel Spring's *Pedagogies of Globalization: The Rise of the Educational Security State* provides a historical and global look at how education has been shaped by racist biopolitical security concerns.

8. Homeless children exist in the United States but tend to be invisible as they are rarely found begging in public spaces. However, increasingly, one can find homeless adolescents congregating in particular areas of certain cities. These street children, like those of the developing world, are viewed as dirty and potentially diseased, if they are viewed at all.

9. Monbiot "Behind the Phosphorus Clouds" ; "US 'Uses Incendiary Arms' in Iraq"; Wilson "US Admits."

10. These experiences characterized our September 2008 family trip to San Diego. My two children were deeply disturbed to see handcuffed people on the side of the freeway and terrified by the low-swooping helicopters with night-penetrating search lights

6 CHILDREN AND THE TWENTY-FIRST CENTURY: RISKY ECONOMIES

1. The SCO includes Russia, China, Kazakhstan, Kyrgyzstan, Tajikistan, and Uzbekistan. India, Pakistan, Mongolia, and Iran have observer-nation status.

2. For discussion of the relationship between poverty and the neoliberal global order backed by U.S. military might, USAID, the World Bank, and the IMF among other powers see Mike Davis's *Planet of Slums* (1–206) and William Engdahl's *Full Spectrum Dominance* (1–228).

3. For an excellent discussion of the challenges facing retirees see Robin Blackburn's *Age Shock: How Finance is Failing Us* (1–321). For a recent article on skyrocketing pension costs for employers stemming from the financial implosion, see Craig Karmin's "Pension Bills to Surge Nationwide" (C1, C2).

WORKS CITED

Abbot, Grace. "Victories for Child Welfare Won in the Last Two Decades." *New York Times,* 10 April 1932: XX4.

Abbott, John J. C. "The Mother at Home." In *Major Problems in the History of American Families and Children,* edited by A. Jabour, 84. New York: Houghton Mifflin, 2005.

Adler, Paul. "New Technologies, New Skills." *California Management Review* 29, no.1 (1986): 9–28.

Aizenman, N.C. "1 in 4 U.S. Kids under 5 Is Latino." *Arizona Republic,* 1 May 2008: A3.

———. "New High in U.S. Prison Numbers Growth Attributed to More Stringent Sentencing Laws." *Washington Post,* 29 February 2008: A1.

———. "Recession Unlikely to Drive away Illegal Immigrants, Report Finds." *Washington Post,* 15 January 2009: A8.

Aizenman, N. C., and Christopher Lee. "U.S. Poverty Rate Drops; Ranks of Uninsured Grow." *Washington Post,* 29 August 2007: A3.

Akinbami, Lara, Jeanne E. Moorman, Paul L. Garbe, and Edward J. Sondik. "Status of Childhood Asthma in the United States, 1980–2007." *Pediatrics* 123. Supplement (March 2009): S131-S145.

Alden, Lyman P. "Non-Sectarian Endowed Child-Saving Institutions." *The History of Child Saving in the US: Report on the Committee on the History of Child-Saving Work.* Boston, MA: Geo. H. Ellis, June 1893. 68–85.

Alini, Erica. "State Income-Tax Revenues Sink." *Wall Street Journal,* 18 June 2009: A4.

Allen, G. E. "Modern Biological Determinism: The Violence Initiative, the Human Genome Project, and the New Eugenics." In *The Practices of Human Genetics,* edited by M. Fortun and E. Mendelsohn, 1–24. Dordrecht: Kluwer, 1999.

"Almost Half of Kids with ADHD Are Not Being Treated." *Science Daily,* 6 August 2006.http://www.sciencedaily.com/releases/2006/08/060804140953.htm (accessed 3 March 2008).

Alonso-Zaldivar, Ricardo. "1 in 5 American Workers Are Uninsured, Study Says." *USA Today,* 24 March 2009. http://www.usatoday.com/money/industries/insurance/2009-03-24-health-insurance_N.htm (accessed 15 June 2009).

Althusser, Louis. "Ideology and Ideological State Apparatuses." In *Lenin and Philosophy and other Essays,* trans. Ben Brewster, 121–76. New York: Monthly Review Press, 1971.

Alvarez, Lizette, and Dan Frosch. "A Focus on Violence by Returning G.I.'s." *New York Times,* 2 January 2009. http://www.nytimes.com/2009/01/02/us/02veterans.html (accessed 2 January 2009).

"America's Children: Key National Indicators of Well-Being, 2007: Lead in the Blood of Children." *Childstats.gov Forum on Child and Family Statistics.* 2007. http://childstats.ed.gov/americaschildren/phenviro3.asp (accessed 3 January 2008).

"America's Fiscal Collapse—Obama's Budget Will Impoverish America with Economist and Author, Michel Chossudovsky." *Guns and Butter KPFA 94.1 FM,* 11 March 2009. http://www.kpfa.org/archive/id/49073 (accessed 4 April 2009).

"America's Longest War." *Economist,* 2 September 2006: 22–24.

American Civil Liberties Union. "ACLU Challenges Defense Department Personnel Policy to Regard Lawful Protests as 'Low-Level Terrorism.' "ACLU. 10 June 2009. http://www.aclu.org/safefree/general/39822prs20090610.html (accessed 24 June 2009).

———. "Military Recruitment Practices Violate International Standards, Says ACLU." ACLU. 13 May 2008. http://www.aclu.org/intlhumanrights/gen/35258prs20080513.html (accessed 9 March 2009).

———. "Soldiers of Misfortune: Abusive U.S. Military Recruitment and Failure to Protect Child Soldiers.13 May 2008. http://www.aclu.org/human-rights/soldiers-misfortune-abusive-us-military-recruitment-and-failure-protect-child-soldiers (accessed 9 March 2009)

———"A Violent Education: Corporal Punishment of Children in U.S. Public Schools." ACLU. February 2009. http://www.aclu.org/pdfs/humanrights/aviolenteducation_execsumm.pdf (accessed 9 March 2009).

Amnesty International. "Children and Human Rights." *Amnesty International* (n.d.). http://www.amnesty.org/en/children (accessed 7 April 2008).

Amparano, J. "Middle-Class Family Lives on the Edge." *Arizona Republic,*17 December 1997: A1, A8.

"An Even Poorer World." *New York Times,* 8 September 2008. http://www.nytimes.com/2008/09/02/opinion/02tue3.html (accessed 8 September 2008).

Andrews, Edmund L. "World Bank Expects Global Economy to Shrink in 2009." *The Herald Tribune,* 9 March 2009: A12.

Annys, A. "Taking Lead Safety into Its Own Hands. *Washington Post,* 10 November 2007: D1.

Ansell, Nicola. *Children, Youth and Development.* Milton Park: Routledge, 2005.

Appel, K. E., and E. A. Strecker. *Practical Examination of Personality and Behavior Disorders: Adults and Children.* New York: The Macmillan Company, 1936.

Apple, Rima. *Perfect Motherhood: Science and Childrearing in America.* New Brunswick, NJ: Rutgers, 2006.

Arehart-Treichel, Joan. "The Bright New World of Brain and Body Scans." *Science News* 109, no. 11 (1976): 170–72.

Ariès, Philippe. *Centuries of Childhood. A Social History of Family Life.* Trans. R. Baldick. New York: Vintage, 1962.

Armstrong, David. *A New History of Identity: A Sociology of Medical Knowledge.* Basingstoke: Palgrave Macmillan, 2002.

———. "The Rise of Surveillance Medicine." *Sociology of Health and Illness* 17 (1995): 393–404.

Armstrong, David. "Children's Use of Psychiatric Drugs Begins to Decelerate. *Wall Street Journal,* 18 May 2009: B1, B10.

"As Workers' Pay Lags, Causes Spur a Debate." *Wall Street Journal,* 31 July 1995: A1.

"Attention Deficit May Be Tied to Smoking, Lead." *Wall Street Journal,* 19 September 2006: D3.

Aylward, Glen P. *Infant and Early Childhood Neuropsychology*. New York: Plenum Press, 1997.

"Back to Genes: More Evidence to Suggest that Genetic Endowment Limits, without Destroying the Possibilities of Social Change." *Economist*, 21 May 1977: 11.

Bair, H. V., and William Herold. "Efficacy of Chlorpromazine in Hyperactive Mentally Retarded Children." *Archives of Neurology and Psychiatry* 74 (1955): 363–64.

Baker, Bernadette. M. *In Perpetual Motion: Theories of Power, Educational History, and the Child*. New York: Peter Lang, 2001.

Baker, Tom, and Jonathan Simon. *Embracing Risk: The Changing Culture of Insurance and Responsibility*. Chicago: University of Chicago Press, 2002.

Banerjee N. "Families Challenging Religious Influence in Delaware Schools." *New York Times*, 29 July 2006. http://www.nytimes.com/2006/07/29/us/29delaware.html?th&emc=th (accessed 7 June 2008).

Banet-Weiser, Sarah. *Kids Rule! Nickelodeon and Consumer Citizenship*. Durham, NC: Duke University Press, 2007.

Barkley, Russell A. "Attention-Deficit Hyperactivity Disorder." *Scientific American* 279, no. 3 (1992): 66–71.

Barnes, Julian. "Military Force, Technology Have Limits, Defense Secretary Robert Gates Warns." *Los Angeles Times*, 30 September 2008. http://www.latimes.com/news/nationworld/nation/la-na-gates30-2008sep30,0,2382416.story (accessed 11 November 2008).

Barnet, Ann, and Richard Barnet. "Childcare Brain Drain?" *Nation*, 12 May 1997: 6–7.

Barry, Dan. "A Boy the Bullies Love to Beat Up, Repeatedly." *New York Times*, 24 March 2008. http://www.nytimes.com/2008/03/24/us/24land.html (accessed 24 March 2008).

Baruchin, Aliyah. "Nature, Nurture and Attention Deficit." *New York Times*, 12 March 2008. http://health.nytimes.com/ref/health/healthguide/esn-adhd-expert.html (accessed 9 March 2007).

Beacon Hill Institute. *BHI Forecast*. 14 December 2008. http://www.beaconhill.org/RevenueForecastsBHI/RF2009–10/PressReleaseBHIMAForecast08–7PM.pdf (accessed 15 December 2008).

Beck, Ulrich. "Living in the World Risk Society." *Economy and Society* 35 (2006): 329–45.

Begley Sharon. "How to Build a Baby's Brain." *Newsweek* 129 (Spring/Summer 1997): 28–32.

———. "Life Events Thwart Scientists' Attempts to Draw DNA Profiles. *Wall Street Journal*, 7 July 2006: B1.

———. "Your Child's Brain." *Newsweek*, 19 February 1996: 55–61.

Bellinger, David C., and Andrew Bellinger, M. "Childhood Lead Poisoning: The Torturous Path from Science to Policy." *Journal of Clinical Investigation* 116, no. 4 (2006): 853–57.

Berger, Meyer. "Help for the Poisoned Child." *Saturday Evening Post*, 16 November 1957: 24–25, 72–80.

Berney, Barbara. "Round and Round It Goes: The Epidemiology of Childhood Lead Poisoning, 1950–1990." *Milbank Quarterly* 71, no. 1 (1993): 3–39.

Bernstein, Elizabeth. "Sending the Baby to a Shrink: Expanding the Field of Infant Mental Health Aims to Head Off Depression and Other Disorders." *Wall Street Journal*, 24 October 2006: D1.

Bernstein, Nina. "Immigrants Challenge Federal Detention System." *New York Times,* 1 May 2008: B3.

Bernstein, Nina, and Julia Preston. "Better Health Care Sought for Detained Immigrants." *New York Times,* 7 May 2008: A18.

Besharov, Douglas, J., Jeffrey Morrow, and Justus Myers. "Costs per Child for Early Childhood Education and Care." *American Enterprise Institute,* 15 March 2007. http://www.aei.org/publications/pubID.26766/pub_detail.asp (accessed 15 March 2007).

Bettelheim, Bruno. *The Empty Fortress: Infantile Autism and the Birth of the Self.* New York: Free Press, 1967.

Bianchi, Suzanne M. "Feminization and Juvenilization of Poverty: Trends, Relative Risks, Causes, and Consequences." *Annual Review of Sociology* 25 (1999): 307–33.

Blackman, B. "A Comparison of Hyperactive and Non-Hyperactive Problem Children." *Smith College Studies in Social Work* 4 (1933): 54–66.

Blank, Marion, and Frances Solomon. "How Shall the Disadvantaged Child Be Taught?" *Child Development* 40, no. 1 (1969): 47–61.

Bloch, Marianne, Devorah Kennedy, Theodora Lightfoot, and Dar Weyenberg, eds. *The Child in the World/ the World in the Child.* New York: Palgrave Macmillan, 2006.

Bloch, Marianne, Kerstin Holmlund, Ingeborg Moqvist, and Thomas Popkewitz, eds. *Governing Children, Families and Education: Restructuring the Welfare State.* New York: Palgrave Macmillan, 2003.

Bourne, Joel K. "The End of Plenty." *National Geographic* 215, no. 6 (2009): 26–59.

Boyden, Jo. "Childhood and the Policy Makers: A Comparative Perspective on the Globalization of Childhood." In *Constructing and Reconstructing Childhood,* edited by Allison James and Alan Prout, 187–226. London: Falmer Press, 1990.

Brace, Charles L. "The Children's Aid Society of New York. Its History, Plans, and Results." In *The History of the Child Saving in the US: Report on the Committee on the History of Child-Saving Work.* Boston, MA: Geo. H. Ellis, 1893. 1–36.

Bradley, Charles. "The Behavior of Children Receiving Benzedrine." *American Journal of Psychiatry* 94 (1937): 577–85.

Bradsher, Keith. "China Slows Purchases of U.S. and Other Bonds." *New York Times,* 13 April 2009. http://www.nytimes.com/2009/04/13/business/global/13yuan.html?th&emc=th (accessed 13 April 2009).

Brancaccio, Marie Teresa. "Educational Hyperactivity: The Historical Emergence of a Concept." *Intercultural Education* 11, no. 2 (2000): 165–77.

Braun, Joe M., Robert S. Kahn, Tanya Froehlich, Peggy Auinger, and Bruce P. Lanphear.

"Exposures to Environmental Toxicants and Attention Deficit Hyperactivity Disorder in U.S. Children." *Environmental Health Perspectives,* 12 December 2006. http://www.ehponline.org/docs/2006/9478/abstract.html (accessed 1 May 2009).

"Breaking Records, Corporate Profits." *Economist,* 12 February 2005. http://www.highbeam.com/doc/1G1–128527130.html (accessed 9 May 2009).

Breggin, Peter R. *Talking Back to Ritalin: What Doctors Aren't Telling You about Stimulants and ADHD.* Cambridge, MA: Perseus, 2001.

Bremner, Robert H. "Other People's Children." *Journal of Social History* 16, no. 3 (1983): 83–103.

Bridges, K. M. B. "Factors Contributing to Juvenile Delinquency." *Journal of the American Institute of Criminal Law and Criminology* 17, no. 4 (1927): 531–80.

Brody, J. E. "Experts Now Link a Learning Disorder to Delinquency." *New York Times,* 13 February 1972: 36.

Brooks, David. "The Cognitive Age." *New York Times,* 2 May 2008. http://www.nytimes.com/2008/05/02/opinion/02brooks.html (accessed 2 May 2008).

Brotherton, J., and L. Gilliver. *The Art of Nursing: Or the Method of Bringing Up Young Children According to the Rules of Physick for the Preservation of Health, and Prolonging Life.* 2nd ed. London: Author, 1733.

Brown, Patricia L. "For 'EcoMoms,' Saving Earth Begins at Home." *New York Times,* 16 February 2008. http://www.nytimes.com/2008/02/16/us/16ecomoms.html *(accessed* 16 February 2008).

Bruer, John. "Brain Science, Brain Fiction." *Educational Leadership* 56, no. 3 (1998): 14–18.

———. "In Search of Brain-Based Education." *Phi-Delta Kappan* 80, no. 9 (1999): 649–57.

Brzezinski, Zbigniew. *The Choice: Global Domination or Global Leadership.* New York: Basic Books, 2004.

———. "The Search for Meaning amid Change." *New York Times,* 6 January 1969: I41.

Buiter, Willem. "The Green Shoots Are Weeds Growing through the Rubble in the Ruins of the Global Economy." *Mavercon,* 8 April 2009. http://blogs.ft.com/maverecon/2009/04/the-green-shoots-are-weeds-growing-through-the-rubble-in-the-ruins-of-the-global-economy/ (accessed 8 April 2009).

"Bullying Behavior: Blame It on Bad Genes?" *ScienceDaily,* 10 March 1999. http://www.sciencedaily.com /releases/1999/03/990310053751.htm (7 May 2008).

Burman, Erica. "Innocents Abroad: Western Fantasies of Childhood and the Iconography of Emergencies." *Disasters* 18, no. 3 (1994): 238–53.

———. "Appealing and Appalling Children." *Psychoanalytic Studies* 1, no .3 (1999): 285–301.

Byers, Randolph. "Lead Poisoning: Review of the Literature and Report on Forty-Five Cases." *Pediatrics* 23 (1959): 585–603.

Byers, Randolph, and Elizabeth Lord. "Late Effects of Lead Poisoning on Mental Development." *American Journal of Diseases in Children* 66 (1943): 471–94.

Calmes, J. "In Bush's 'Ownership Society,' Citizens Would Take More Risk." *Wall Street Journal,* 28 February 2005: A1, A12.

Campbell, David. *Writing Security: United States Foreign Policy and the Politics of Identity.* Minneapolis: University of Minnesota Press, 1998.

Cantwell, D. P. "Psychiatric Illness in the Families of Hyperactive Children." *Archives of General Psychiatry* 27 (1972): 414–17.

Caplan, Frank. *The First Twelve Months of Life.* New York: Putnam, 1971.

Carey, Benedict. "Bipolar Illness Soars as a Diagnosis for the Young." *New York Times,* 4 September 2007. http://www.nytimes.com/2007/09/04/health/04psych.html (accessed 4 September 2007).

———. "Parenting as Therapy for Child's Mental Disorders." *New York Times* 22 December 2006. http://query.nytimes.com/gst/fullpage.html?sec=health&res=9506E7DC1131F931A15751C1A9609C8B63 (accessed 22 December 2006).

Carson, Rachel. *Silent Spring.* Boston, MA: Houghton Mifflin, 1962.

"Cash-Hungry Kids." *Business Week,* 7 July 2008: 13.

"Cash-Machine." *Atlantic Monthly,* 4 May 2009: 58–59.

Caspi, A. "Social Selection, Social Causation, and Developmental Pathways: Empirical Strategies for Better Understanding How Individuals and Environments are Linked across the Life-Course." In *Paths to Successful Development: Personality in the Life Course*, edited by L. Pulkkinen and A. Caspi, 281–301. New York: Cambridge University Press, 2002.

Caspi, Avshalom, Joseph McClay, Terrie E. Moffitt, Jonathan Mill, Judy Martin, Ian W. Craig, Alan Taylor, and Richie Poulton. "Role of Genotype in the Cycle of Violence in Maltreated Children." *Science* 297 (2002): 851–54.

Castel, Robert. "From Dangerousness to Risk." In *The Foucault Effect: Studies in Governmentality*, edited by Graham Burchell, Colin Gordon, and Peter Miller, 281–98. Chicago: University of Chicago Press, 1991.

Cauchon, Dennis. "Employed See Tough Times Too. *USA Today,* 12 June 2009. http://www.usatoday.com/money/economy/employment/2009-06-11-workweek_N.htm (accessed 15 June 2009).

"CEO Council on Health Care." *Wall Street Journal,* 23 November 2008. http://blogs.wsj.com/ceo-council/2008/11/23/health-care/ (accessed 23 November 2008).

Chaker, A. M. "In Obesity Wars, A New Backlash." *Wall Street Journal,* 14–15 April 2007: A1, A8.

Chappell, Ben. "Rehearsals of the Sovereign: States of Exception and Threat Governmentality." *Cultural Dynamics* 18 (2006): 313–34.

Childers, A. T. "Hyper-activity in Children Having Behavior Disorders." *American Journal of Orthopsychiatry* 5 (1935): 227–43.

"Children from Happy Homes Have Less Trouble in School." *Science News Letter,* 4 June 1932: 352.

Cho, David, and Neil Irwin. "Financial Rescue Turns to Toxic Assets." *Washington Post,* 4 March 2009: D1.

Clements, Sam D., and John E. Peters. "Minimal Brain Dysfunction in the School Age Child: Diagnosis and Treatment." *Archives of General Psychiatry* 6, no. 3 (1962): 185–97.

Cohen, David B, and John W. Wells. *America's National Security and Civil Liberties in an Era of Terrorism.* New York: Palgrave Macmillan, 2004.

Cohn, D'Vera. "Parents' Top Goal: Thinkers." *Arizona Republic,* 28 November 1999: A23.

Cole, Robert. *The Story of Ruby Bridges.* New York: Scholastic, 1995.

Colen, B. D. "Study Links Level of Lead in Blood to IQ Test Performance." *Washington Post,* 6 May 1978: A2.

Collins, James. "The Day-Care Dilemma." *Time,* 3 February 1997: 58–62.

Conger, R. D., K. J. Conger, G. H. Elder, F. O. Lorenz, and R. L. Simons. "Economic Stress, Coercive Family Process, and Developmental Problems of Adolescents." *Child Development* 65 (1994): 541–61.

Conrad, Peter, and Deborah *Potter.*"From Hyperactive Children to ADHD Adults: Observations on the. Expansion of Medical Categories." *Social Problems* 47 (2000): 559–82.

Conrad, Peter, and Joseph W. Schneider. *Deviance and Medicalization.* Philadelphia: Temple University Press, 1992.

Constable, Pamela. "Americans, Europeans Share Immigration Worries." *Washington Post,* 18 November 2008: A15.

Cooke, R. "Possible Link of Violence, Gene Found." *Arizona Republic,* 4 August 2002: A18.

Cooper, Melinda. *Life as Surplus: Biotechnology and Capitalism in the Neoliberal Era*. Seattle: University of Washington Press, 2008.

Cooper, Sarah B. "The Kindergarten in Its Bearing upon Crime, Pauperism, and Insanity." In *The History of Child Saving in the US: Report on the Committee on the History of Child-Saving Work*. Boston, MA: Geo. H. Ellis, 1893. 86–98.

Corn, Jacqueline K. "Historical Perspectives to a Current Controversy on the Clinical Spectrum of Plumbism." *The Milbank Memorial Fund Quarterly. Health and Society* 53, no. 1 (1975): 93–114.

Cotton, John W., and Robert H. Ellis. "The Transistorized Child." *PsycCritiques* 19, no. 1 (1974): 37–38.

Cox, Roger. *Shaping Childhood: Themes of Uncertainty in the History of Adult-Child Relationships*. London: Verso, 1996.

Cravens, Hamilton. "Child Saving in Modern America 1870s-1990s." In *Children at Risk in America: History, Concepts, Public Policy*, edited by Roberta Wollons, 3–31. Albany: SUNY, 1993.

Crewe, Emma, and Elizabeth Harrison. *Whose Development? An Ethnography of Aid*. New York: St. Martin's Press, 1998.

Crinson, Matthew. *Building the Invisible Orphanage*. Cambridge, MA: Harvard University Press, 1998.

Cross, Gary. *Kid's Stuff: Toys and the Changing World of American Childhood*. Cambridge, MA: Harvard University Press, 1997.

Cummings, Scott L. "Community Economic Development as Progressive Politics: Toward a Grassroots Movement for Economic Justice." *Stanford Law Review* 54, no. 3 (2001): 399–493.

Cuordileone, K. A. *Manhood and the American Political Culture in the Cold War*. Routledge New York, 2005.

Dash, Eric, and Andrew Martin. "Banks Brace for Credit Card Write-Offs." *New York Times* 11 May 2009. http://www.nytimes.com/2009/05/11/business/11credit.html?_r=1&th&emc=th (accessed 11 May 2009).

Davies, Dean F., and Alice H. Davies. "Lung Cancer: Cigarette Smoking as a Cause." *American Journal of Nursing* 61, no. 4 (1961): 65–69.

Davis, A. "Call Goes out to Rein in Grain Speculators." *Wall Street Journal*, 22 April 2008: A4.

Davis, Mike. *Dead Cities: and Other Tales*. New York: New Press, 2003.

———. *Planet of Slums*. London: Verso, 2006.

Deacon, Roger, A. *Fabricating Foucault: Rationalising the Management of Individuals*. Milwaukee, WI: Marquette University Press, 2003.

———. "Moral Orthopedics: A Foucauldian Account of Schooling as Discipline." *Telos* 130 (2005): 84–102.

———. "Truth, Power and Pedagogy: Michel Foucault on the Rise of the Disciplines." *Educational Philosophy and Theory* 34, no. 4 (2002): 435–58.

Dean, Mitchell. *The Constitution of Poverty: Toward a Genealogy of Liberal Governance*. London: Routledge, 1990.

———. *Governing Societies*. New York: McGraw-Hill, 2007.

———. *Governmentality: Power and Rule in Modern Society*. London: Sage, 1999.

———. "Liberal Government and Authoritarianism." *Economy and Society* 31 (2002): 37–61.

"Death by Detention." *New York Times*, 6 May 2008: A26.

DeGrandpre, Richard. *Ritalin Nation: Rapid-Fire Culture and the Transformation of Human Consciousness*. New York: W.W. Norton, 1999.

DeLong, Brad. "Five Points to Eliminate Confusion about the U.S. Macroeconomy." *Seeking Alpha*, 19 June 2009. http://seekingalpha.com/article/144149-five-points-to-eliminate-confusion-about-the-u-s-macroeconomy (accessed 19 June 2009).

DeMause, Lloyd. "The Evolution of Childhood." In *The History of Childhood*, edited by Lloyd DeMause, 1–74. New York: The Psychohistory Press, 1974.

DePalma, Anthony. "15 Years on the Bottom Rung." *New York Times*, 26 May 2005: A1.

DeParle, Jason. "A Global Trek to Poor Nations, from Poorer Ones." *New York Times*, 27 December 2007. http://www.nytimes.com/2007/12/27/world/americas/27migration.html?_r=1&pagewanted=3 (accessed 27 December 2007).

———. "Slumping Economy Tests Aid System Tied to Jobs." *New York Times*, 1 June 2009. http://www.nytimes.com/2009/06/01/us/politics/01poverty.html?_r=1&th&emc=th (accessed 1 June 2009).

Derose, K. P., J. J. Escarce, and N. Lurie. "Immigrants and Health Care: Sources of Vulnerability." *Health Affairs* 26, no. 5 (September–October 2007): 1258–68.

Dewan, Shaila. "Many Children Lack Stability Long after Storm." *New York Times*, 6 December 2008: A1.

Dickerson, Marla. "Placing Blame for Mexico's Ills." *Los Angeles Times*, 1 July 2006: C1.

Dillon, Sam. "Hard Times Hitting Students and Schools." *New York Times*, 1 September 2008. http://www.nytimes.com/2008/09/01/education/01school.html *(accessed* 1 September 2008).

DiPietro, Janet. "Baby and the Brain: Advances in Child Development." *Annual Review of Public Health* 21 (2000): 455–71.

Doering, John A. "Trends in Mental Hygiene. *Delaware State Medical Journal* 19 (1947): 87–90.

Donatelli, Rosemary. "Will Your Child Fail in School?" *Saturday Evening Post*, 27 October 1956: 86, 89–90.

Donzelot, J. "Michel Foucault and Liberal Intelligence." *Economy and Society* 37 (2008): 115–34.

———. *The Policing of Families.* Trans. R. Hurley. Baltimore: Johns Hopkins University Press, 1977.

Dorey, Annette, K. V. *Better Baby Contests: The Scientific Quest for Perfect Childhood Health in the Early Twentieth Century.* Jefferson, NC: McFarland & Co., 1999.

Douglas, Wilder, L. "To Save the Black Family, the Young Must Abstain." *Wall Street Journal*, 28 March 1991: A14.

Dreazen, Yochi, J. "U.S. News: Wars Harming Mental Health of Soldiers, Spouses; Problems Present Long, Hidden Toll; Help Often Avoided." *Wall Street Journal*, 8 April 2008: A4.

Dubber, Markus D., and Mariana Valverde, eds. *The New Police Science; The Police Power in Domestic and International Governance.* Stanford, CA: Stanford University Press, 2006.

Duffield, Mark. *Global Governance and the New Wars: The Merging of Development and Security.* London: Zed Books, 2001.

DuGay, Paul. "Organizing Identity: Entrepreneurial Governance and Public Management." In *Questions of Cultural Identity*, edited by S. Hall and P. DuGay, 151–69. London: Sage, 1996.

Dumit, Joseph, and Robbie Davis-Floyd. "Introduction." In *Cyborg Babies: From Techno-Sex to Techno-Tots*, edited by R. Davis-Floyd and J. Dumit, 1–20. New York: Routledge, 1998.

Eckholm, Erik. "In Turnabout, Infant Deaths Climb in South." *New York Times*, 22 April 2007. http://www.nytimes.com/2007/04/22/health/22infant.html?t h=&emc=th&pagewanted=print (accessed 22 April 2007).

———. "Murders by Black Teenagers Rise, Bucking a Trend." *New York Times*, 29 December 2008. http://www.nytimes.com/2008/12/29/us/29homicide.html *(accessed* 29 December 2008).

———. "Safety Net Is Fraying for the Very Poor." *New York Times*, 5 July 2009. http://www.nytimes.com/2009/07/05/us/05safetynet.html?_r=1&th&emc=th (accessed 5 July 2009).

———. "Working Poor and Young Hit Hard in Downturn." *New York Times*, 9 November 2008. http://www.cfed.org/imageManager/_documents/NYT_11–08-08.pdf *(accessed* 22 June 2009).

"Economic Crisis Set to Drive 53 Million More People into Poverty in 2009—World Bank." *United Nations News Center*, 13 February 2009. http://www.un.org/apps/news/story.asp?NewsID=29897&Cr=financial&Cr1=crisis (accessed 10 May 2009).

Edwards, Michael. *Just Another Emperor? The Myths and Realities of Philanthrocapitalism*. New York: Demos, 2008.

Egan, Matt. "Forget Nukes—Watch Out for Economic War." *Fox News*, 24 April 2009. http://www.foxbusiness.com/story/markets/economy/economic-war-poses-threat-recession/ (accessed 9 July 2009).

Einhorn, David. "Private Profits and Socialized Risk." Grant's Investment Conference, 8 April 2008. http://mrmortgage.typepad.com/blog/files/david_einhorn_private_profits_socialized_risk_40808.pdf (accessed 6 June 2009).

Eilperin, Juliet. "EPA Will Mandate Tests on Pesticide Chemicals" *Washington Post*, 16 April 2009: A1

Ehrenreich, Barbara. *Fear of Falling: The Inner Life of the Middle Class*. New York: Pantheon, 1989.

Ehrenreich, Barbara, and Deirdre English. *For Her Own Good: 150 Years of the Expert's Advice to Women*. Garden City, NY: Anchor Press, 1978.

Eisenberg, Arlene. *What to Expect the First Year*. New York: Workman, 1989.

Eisenberg, Arlene, Heidi Murkoff, and Sandee Hathaway. *What to Expect the Toddler Years*. New York: Workman, 1996.

Elyachar, Julia. *Markets of Dispossession: NGOs, Economic Development, and the State in Cairo*. Durham, NC: Duke University Press, 2005.

Engdahl, F. William. *Full Spectrum Dominance: Totalitarian Democracy in the New World Order*. Baton Rouge, LA: Third Millennium Press, 2009.

English, Peter. *Old Paint: A Medical History of Childhood Lead-Paint Poisoning in the United States to 1980*. New Brunswick, NJ: Rutgers University Press, 2001.

Ericson, Richard, V., and Aaron Doyle. "Risk and Morality." In *Risk and Morality*, edited by Richard V. Ericson and Aaron Doyle, 1–10. Toronto: University of Toronto Press, 2003.

Escobar, Arturo. *Encountering Development: The Making and Unmaking of the Third World*. Princeton, NJ: Princeton University Press, 1995.

Escobar, Pepe. "Pipeline-Istan: Everything You Need to Know about Oil, Gas, Russia, China, Iran, Afghanistan and Obama." *Alternet*, 14 May 2009. http://www.alternet.org/story/139983 (accessed 14 May 2009).

Evans, Kelly. "Jobless Rate Tops 8%, Highest in 26 Years." *Wall Street Journal*, 7–8 March 2009: A1, A2.

Evans, Kelly. "Ranks of Older Workers Swell as Losses Shorten Retirement. *Wall Street Journal,* 9–10 May 2009: A2.

Ewen, Stuart. *All Consuming Images: The Politics of Style in Contemporary Culture.* Rev. ed. New York: HarperOne, 1990.

"Facts on Policy: Consumer Spending." *Hoover Institution,* 19 December 2006. http://www.hoover.org/research/factsonpolicy/facts/4931661.html (accessed 17 May 2009).

Fallows, James. "Dr. Doom Has Some Good News." *Atlantic Monthly,* July/August 2009: 88–90, 91.

Fantz, Ashley. "Children Forced into Cell-Like School Seclusion Rooms." *CNN,* 17 December 2008. http://www.cnn.com/2008/US/12/17/seclusion.rooms/index.html (accessed 17 December 2008).

Favole, Jared D. "FDA Cites Limitations of ADHD Drug Study." *Wall Street Journal,* 16 June 2009: D6.

Federal Interagency Forum on Child and Family Statistics. *America's Children: Key National Indicators of Well-Being, 2007.* Washington, D.C.: U.S. Government Printing Office, 2007.

Fee, Elizabeth. "Public Health in Practice: An Early Confrontation with the 'Silent Epidemic' of Childhood Lead Poisoning." *Journal of the History of Medicine and Allied Sciences* 45 (1990): 570–606.

Ferguson, H. "Cleveland in History: the Abused Child and Child Protection, 1880–1914. 1890–1930." In *In the Name of the Child: Health and Welfare, 1880–1940,* edited by R. Cooter, 146–73. London: Routledge, 1992. "Financial Crisis to Deepen Extreme Poverty, Increase Child Mortality Rates—UN Report." *United Nations News Center,* 3 March 2009. http://www.un.org/apps/news/story.asp?NewsID=30070&Cr=Financial+crisis&Cr1 (accessed 3 March 2009).

Finn, Janet L. "Troubled in Paradise: A Critical Reflection Youth, Trouble, and Intervention." In *Disciplining the Child Via the Discourse of the Professions,* edited by Roxanna P. Transit, 90–129. Springfield, IL: Charles C. Thomas, 2004.

First Focus. "The Cost of Doing Nothing." *First Focus: Making Children & Families the Priority,* 16 December 2008. http://www.firstfocus.net/pages/3534/ (accessed 19 January 2009).

Fisher, Marc. "Don't Gum Up Sex-Ed; Leave Instruction to Professional Teachers." *Washington Post,* 15 February 2007: B1.

Fitch, Ed. "Baby Boomlet Builds Blockbuster Sales." *Advertising Age* 56, no. 12 (1985): 15, 35.

Fitzsimons, Patrick "Neoliberalism and Education: The Autonomous Chooser." *Radical Pedagogy* 4 no. 2 (2002): http://radicalpedagogy.icaap.org/content/issue4_2/04_fitzsimons.html (accessed 5 March 2007).

Flattau, Pamela Ebert, Jerome Bracken, Richard Van Atta, Ayeh Bandeh-Ahmadi, Rodolfo de la Cruz, and Kay Sullivan. "The National Defense Education Act of 1958: Selected Outcomes." IDA: Science and Technology Policy Institute, March 2006. http://www.ida.org/stpi/pages/D3306-FINAL.pdf (accessed 19 May 2009).

Fletcher, Michael A. "1 in 4 Working Families Now Low-Wage, Report Finds." *Washington Post,* 15 October 2008: D3.

———. "Middle-Class Dream Eludes African American Families." *Washington Post,* 13 November 2007: A1.

Ford, William W. "The Present Status and the Future of Hygiene or Public Health in America." *Science, New Series* 42, no. 1070 (1915): 1–13.

Foster, John B., and Fred Magdoff. *The Great Financial Crisis: Causes and Consequences*. New York: Monthly Review Press, 2009.

Foucault, Michel. *The Birth of Biopolitics: Lectures at the Collège de France 1978–1979*. Edited by Michel Senellart. Translated by G. Burchell. Basingstoke: Palgrave Macmillan, 2008.

———. *Discipline and Punish*. Translated by A. Sheridan. New York: Vintage Books, 1979.

———. "Governmentality." In *The Foucault Effect*, edited by G. Burchell, C. Gordon, and P. Miller, 97–104. London: Harvester Wheatsheaf, 1991.

———. *The History of Sexuality: An Introduction*. Vol 1. Translated by Robert Hurley. New York: Vintage, 1990.

———. *The Order of Things: An Archeology of the Human Sciences*. New York: Vintage, 1994.

———. *Power/Knowledge: Selected Interviews and Other Writings*. Translated by C. Gordon. Brighton, UK: Harvester, 1977.

———. *Psychiatric Power: Lectures at the Collège de France 1973–1974*. Edited by J. Lagrange. Translated by G. Burchell. Basingstoke: Palgrave Macmillan, 2006.

———. *Security, Territory, Population: Lectures at the Collège de France 1977–78*. Edited by M. Senellart. Translated by G. Burchell. Basingstoke: Palgrave Macmillan, 2007.

———. *Society Must Be Defended: Lectures at the Collège de France 1975–1976*. Edited by M. Bertani and A. Fontana. Translated by D. Macey. New York: Picador, 2003.

———. "The Subject and Power." In *Michel Foucault: Beyond Structuralism and Hermeneutics*, edited by H. L. Dreyfus and P. Rabinow, 208–64. Chicago: University of Chicago Press, 1983.

———. "Technologies of the Self." In *Technologies of the Self: A Seminar with Michel Foucault*, edited by L. H. Martin, H. Gutman, and P. H. Hutton, 16–49. Amherst: University of Massachusetts Press, 1988.

Fraser, Christian. "New Evidence of Gaza Child Deaths." *BBC News*, 22 January 2009. http://news.bbc.co.uk/2/hi/programmes/from_our_own_correspondent/7843307.stm (accessed 22 January 2009).

Frazier, Claude A. "Suffer Little Children." *Saturday Evening Post*, October (1980): 72–77.

Freeman, Roger D. "Minimal Brain Dysfunction, Hyperactivity, and Learning Disorders: Epidemic or Episode?" *The School Review* 85, no. 1 (1976): 5–30.

Freier, Nathan. "Known Unknowns: Unconventional 'Strategic Shocks' in Defense Strategy Development." *Strategic Studies Institute*, November 2008. http://www.StrategicStudiesInstitute.army.mil/ (accessed 12 December 2008).

Frickel, Scott. "When Convention Becomes Contentious: Organizing Science Activism in Genetic Toxicology." In *New Political Sociology of Science: Institutions, Networks, and Power*, edited by Scott Frickel and Kelly Moor, 185–210. Madison: University of Wisconsin Press, 2006.

Friedman, Milton. *Capitalism and Freedom*. Chicago: University of Chicago Press, 1962.

Friend, Angela, John C. DeFries, and Richard Olson. "Parental Education Moderates Genetic Influences on Reading Disability." *Psychological Science* 19, no. 11 (2008): 1124–30. http://www.psychologicalscience.org/journals/ps/19_11_inpress/Friend.pdf (accessed 1 June 2009).

Frum David. "The Vanishing Republican Voter." *New York Times Magazine*, 7 September 2008. http://www.nytimes.com/2008/09/07/magazine/07Inequality-t.html (accessed 7 September 2008).

Galaburda, Albert, Marjorie LeMay, Thomas Kemper, and Norman Geschwind. "Right-Left Asymmetries in the Brain." *Science, New Series* 199, no. 4331 (February 1978): 852–56.

Gallagher, J., and R. Clifford. "The Missing Support Infrastructure in Early Childhood." *Early Childhood Research and Practice* 2, no. 1 (2000). http://ecrp.uiuc.edu/v2n1/gallagher.html (accessed 17 March 2009).

Gallup Poll. "Gallup's Pulse of Democracy: Education." *Gallup*, n.d. http://www.gallup.com/poll/1612/Education.aspx (accessed 7 May 2009).

———. "Gallup's Pulse of Democracy: Religion." *Gallup*, n.d. http://www.gallup.com/poll/1690/Religion.aspx (accessed 7 May 2009).

Gamboa, Suzanne. "Report: 100,000 Deportees had U.S. Children." *Arizona Republic*, 14 February 2009: A21.

Garfield, Gail. "Hurricane Katrina: The Making of Unworthy Disaster Victims." *Journal of African American Studies* 10, no. 4 (2007): 55–74.

Garza, Cynthia L. "Spare the Rod? Not at Many Public Schools." *The Times-Union*, 12 January 2004. http://www.nospank.net/n-l26r.htm (accessed 9 February 2009).

"Gaza Humanitarian Situation." *BBC News*, 30 January 2009. http://news.bbc.co.uk/2/hi/middle_east/7845428.stm (accessed 20 January 2009).

Gibson, V. M. "Employer Support for Nursing Mothers Yields Numerous Benefits." *HR Focus* 70, no. 9 (September 1993): 17.

Gilman, Charlotte P. *Concerning Children*. Walnut Creek, CA: Rowman & Littlefield, 2003.

Giroux, Henry. "Authoritarianism's Footprint and the War against Youth." *Dissident Voice*, December 2003. http://www.dissidentvoice.org/Articles9/Giroux_War-On-Youth.htm (accessed 7 April 2009).

———. "Beyond the Biopolitics of Disposability: Rethinking Neoliberalism in the New Gilded Age." *Social Identities* 14, no. 5 (2008): 587–620.

———. "The New Authoritarianism in the United States." *Dissident Voice*, January 2006. http://www.dissidentvoice.org/Jan06/Giroux-3.htm (accessed 7 April 2009).

Glaberson, William. "A Legal Debate in Guantánamo on Boy Fighters." *New York Times*, 3 June 2007: A1.

Glick, D. "Rooting for Intelligence." *Newsweek*, special issue, March 22 1997, 32.

Goddard, Jacquie. "Beating at Bootcamp is Blamed for Boy's Death." *Times Online*, 16 March 2006. http://www.timesonline.co.uk/tol/news/world/us_and_americas/article741657.ece (accessed 12 January 2009).

Goldman-Rakic, Patricia S. "Development of Cortical Circuitry and Cognitive Function." *Child Development* 58, no. 3 (1987): 601–22.

Goodman, Peter S. "Crisis Sends Dollars Flowing Back to U.S." *Herald Tribune*, 19 March 2009: A1, A15.

Goodnough, A. "Census Shows a Modest Rise in U.S. Income." *New York Times*, 29 August 2007. http://www.nytimes.com/2007/08/29/us/29census.html?fta=y (accessed 29 August 2007).

Gopnik, Alison, Andrew Meltzoff, and Patria Kuhl. *The Scientist in the Crib: Minds, Brains, and How Children Learn*. New York: William Morrow, 1999.

Gorman, Anna. "Immigrant Detention Facility is Considered; Contractors Are Sought for a Possible Project in the L.A. Area." *Los Angeles Times,* 3 February 2009: B5.

Gottweis, Herbert. *Governing Molecules: The Discursive Politics of Genetic Engineering in Europe and the United States.* Cambridge: MIT Press, 1998.

Goulet, L. R. "Optimum Learning and Optimal Environments: A View from a Child Development Perspective." *Educational Technology* 11, no. 2 (1971): 13–18.

Graham, C., and D. Neu. "Standardized Testing and the Construction of Governable Persons." *Journal of Curriculum Studies* 36 (2004): 295–319.

Graham, L. "Beyond Manipulation: Lillian Gilbreth's Industrial Psychology and the Governability of Women Consumers." *Sociological Quarterly* 38, no. 4 (1997): 539–65.

Grandjean, P., and P. J. Landrigan. "Developmental Neurotoxicity of Industrial Chemicals." *Lancet* 368, no. 9553 (2006): 2167–78.

Grant, Julia. *Raising Baby by the Book: The Education of American Mothers.* New Haven, CT: Yale University Press, 1998.

Greene, Ron W. *Malthusian Worlds: U.S. Leadership and the Governing of the Population Crisis.* Boulder, CO: Westview, 1999.

Greven, P. "The Protestant Temperament: Patterns of Child-Rearing, Religious Experience, and the Self in Early America." In *Major Problems in the History of American Families and Children,* edited by A. Jabour, 86–96. New York: Houghton Mifflin, 2005.

Grossberg, Larry. *Caught in the Crossfire: Kids, Politics, and America's Future.* Boulder, CO: Paradigm, 2005.

Grossberg, Michael. *Governing the Hearth: Law and the Family in Nineteenth Century America.* Chapel Hill: University of North Carolina Press, 1985.

Gubernick, Lisa, and Marla Matzer. "Babies as Dolls." *Forbes,* 27 February 1995: 78, 82.

Gunnar, M. R. "Quality of Early Care and Buffering of Neuroendocrine Stress Reactions: Potential Effects on the Developing Human Brain." *Preventive Medicine* 27(1998): 208–11.

Habermas, Jurgen. *The Structural Transformation of the Public Sphere: An Inquiry into a Category of Bourgeois Society.* Translated by Thomas Berger. Cambridge: MIT Press, 1991.

Hacker, Jacob. *The Great Risk Shift.* Oxford: Oxford University Press, 2006.

Haenszel, William. "Epidemiological Tests of Theories on Lung Cancer Etiology." *Public Health Reports* 71, no. 2 (1956): 163–72.

Haith, M. M., and J. Campos. "Human Infancy." *Annual Review of Psychology* 28 (1977): 251–93.

Harden, B. "A City's Changing Face; Wealth, Race Guiding Which New Orlenians Stay, and Which Never Return." *Washington Post,* 17 May 2006: A1.

———. "High-Tech Revolution in Robotics Under Way; Unlimited Uses Could Transform Way People Work." *Washington Post,* 19 December 1982: H1.

Harmon, Amy. "That Wild Streak? Maybe It Runs in the Family." *New York Times,* 15 June 2006. http://www.nytimes.com/2006/06/15/health/15gene.html (15 June 2006).

Harris, Gardiner. "Proof Is Scant on Psychiatric Drug Mix for Young." *New York Times,* 23 November 2006: A1.

Harris, Gardiner. "Research Center Tied to Drug Company." *New York Times,* 24 November 2008. http://www.nytimes.com/2008/11/25/health/25psych.html (accessed 24 November 2008).

———. "Use of Antipsychotics in Children is Criticized." *New York Times,* 19 November 2008. http://www.nytimes.com/2008/11/19/health/policy/19fda. html *(accessed* 19 November 2008).

Harris, Gardiner, and Carey, Benedict. "Researchers Fail to Reveal Full Drug Pay." *New York Times,* 8 June 2008. http://www.nytimes.com/2008/06/08/ us/08conflict.html *(accessed* 8 June 2008).

Harvey, David. *A Brief History of Neoliberalism.* Oxford: Oxford University Press, 2005.

Harwood, Valerie. *Diagnosing "Disorderly" Children: A Critique of Behaviour Disorder Discourses.* London: Routledge, 2006.

Hayek, Friedrich A. von. *The Road to Serfdom.* Chicago: University of Chicago Press, 1944.

Healy, David. *The Creation of Psychopharmacology.* Cambridge, MA: Harvard University Press, 2002.

———. "Psychopharmacology and the Government of the Self." *Academy for the Study of Psychoanalytic* Arts, n.d. http://www.academyanalyticarts.org/healy. htm **(accessed** 10 June 2007).

Healy, William. *The Individual Delinquent; A Text-Book of Diagnosis and Prognosis for All Concerned in Understanding Offenders.* Boston, MA: Little, Brown, 1915.

Hendershot, Heather. *Nickelodeon Nation: The History, Politics and Economics of America's Only Television Channel for Kids.* Albany: New York University Press, 2004.

Henderson, Nell. "Effect of Immigration on Jobs, Wages is Difficult for Economists to Nail Down." *Washington Post,* 15 April 2006: D1.

Herbert, Bob. "6-Year-Olds Under Arrest." *New York Times,* 9 April 2007: A17.

———. "Harassed in the Classroom." *New York Times,* 3 July 2007: A17.

Hernandez, Kathleen, and George Tokuhata. "Epidemiological Study of Childhood Leukemia in Memphis and Shelby County, 1939–1962. *Public Health* 81, no. 7 (July 1966): 598–606.

Herrnstein, R. J., and C. Murray. *The Bell Curve: Intelligence and Class Structure in American Life.* New York: Free Press, 1994.

Hill, David. "Educational Perversion and Global Neo-Liberalism: A Marxist Critique." *Cultural Logics,* October 2004. http://eserver.org/clogic/2004/hill. html (accessed 22 October 2004).

"History of Social Security Related Legislation." *Social Security Online.* http:// www.ssa.gov/legislation/history/ (accessed 7 March 2009).

Hochschild, Arlie Russell. *The Commercialization of Intimate Life: Notes from Work and Home.* Berkeley: University of California Press, 2003.

Hogan, Michael. *Cross of Iron: Harry S. Truman and the Origins of the National Security State 1945–1954.* Cambridge: Cambridge University Press, 2000.

Hookway, J., and L. Etter. "Rice Profiteers Draw Asian Clampdowns." *Wall Street Journal,* 18 April 2008: A7.

Hossein-Zadeh, Ismael. *The Political Economy of U.S. Militarism.* New York: Palgrave Macmillan, 2007.

Hsu, Spencer S. "Immigration Prosecutions Hit New High: Critics Say Increased Use of Criminal Charges Strains System." *Washington Post,* 2 June 2008: A1.

Hsu, Spencer S. "No Phone Calls for Many Detainees; GAO Report Cites Violations of Guidelines for Dealing with Immigrants." *Washington Post,* 13 July 2007: A2.

Hsu, Spencer S., and Ann S. Tyson. "Pentagon to Detail Troops to Bolster Domestic Security." *Washington Post,* 1 December 2008: A1.

Hudson, Michael. "Financial Bailout: America's Own Kleptocracy: The Largest Transformation of America's Financial System since the Great Depression." *Global Research,* 20 September 2008. http://www.globalresearch.ca/index. php?context=va&aid=10279 (accessed 20 September 2008).

Huebner, Albert. L. "Childhood's Hidden Epidemic." *Nation,* 4 March 1978: 242–44.

Hulbert, Ann. *Raising America: Experts, Parents, and a Century of Advice about Children.* New York: Vintage, 2004.

Hultqvist, Kenneth, and Gunilla Dahlberg. *Governing the Child in the New Millennium.* New York: Routledge Falmer, 2001.

Huntington, Samuel P. "The Clash of Civilizations?" *Foreign Affairs* 72, no. 3 (1993): 22–49.

———. "Politics in a World of Hybrid Cultures: Migration is the Central Issue of the 21st Century." *New Perspectives Quarterly* 18, no. 2 (2001): 22–24.

Hursh, David. "Neoliberalism and the Control of Teachers, Students, and Learning: The Rise of Standards, Standardization, and Accountability." n.d. http://eserver. org/clogic/4–1/hursh.html (accessed 22 October 2004).

Hwang, Suein. "Anxiety High: Moving for Schools." *Wall Street Journal,* 2 January 2007: A1.

"HyperactivityandAcademicAchievementCouldBeLinkedByGenetics." *ScienceDaily,* 17 May 2007. http://www.sciencedaily.com/releases/2007/05/070517063103. htm (accessed 2 March 2008).

International Monetary Fund. "World Economic Outlook (WEO): Crisis and Recovery." April 2009. http://www.imf.org/external/pubs/ft/weo/2009/01/ index.htm (accessed 1 May 2009).

"Increase in Unemployed Veterans." *U.S. News and World Report,* 18 November 2008. http://www.usnews.com/articles/news/2008/11/18/hot-docs-veter- ans-unemployment-rises-a-cultural-benefit-from-immigration.html (accessed 19 March 2009).

"IMF: World Economy Won't Recover until 2010." *CNN.com/World Business 2009,* 23 April 2009. http://edition.cnn.com/2009/BUSINESS/04/22/imf.fore- cast/index.html (accessed 23 April 2009).

Insel, T. R., and F. S. Collins. "Psychiatry in the Genomics Era." *American Journal of Psychiatry* 160 (2003): 616–20.

Ip, Greg. "Income-Inequality Gap Widens." *Wall Street Journal,* 12 October 2007: A3.

"Iraq Children 'Paying High Price.'" *BBC News,* 21 December 2007. http://news. bbc.co.uk/2/hi/middle_east/7156399.stm (accessed 13 February 2008).

Isaacs, S. L., and S. A. Schroeder. "Class—The Ignored Determinant of the Nation's Health." *New England Journal of Medicine* 351, no. 11 (2004): 1137–42.

Jackson, Kenneth T. *Crabgrass Frontier: The Suburbanization of the United States.* New York: Oxford University Press, 1985.

Jenkins, Chris L. "Homelessness: The Family Portrait." *Washington Post,* 16 February 2009: A1.

Jennings, H. S. *The Biological Basis of Human Nature.* New York: W. W. Norton, 1930.

Jensen, Arthur. "How Much Can We Boost I.Q. and Scholastic Achievement?" *Harvard Educational Review* 33 (1969): 1–123.

Jesella, K. "Mom's Mad. And She's Organized." *New York Times,* 22 February 2007. http://www.nytimes.com/2007/02/22/fashion/22mothers.html?th&emc=th (accessed 22 February 2007).

"Job Losses Hit U.S. Latino Immigrants Hard." *Reuters,* 12 February 2009. http://www.reuters.com/article/domesticNews/idUSTRE51B69C20090212 (accessed 3 March 2009).

John, Mary. "Children's Rights in a Free Market Culture." In *Children and the Politics of Culture,* edited by Sharon Stephens, 105–37. Princeton: Princeton University Press, 1995.

Johnson, Chalmers. *The Sorrows of Empire: Militarism, Secrecy, and the End of the Republic.* New York: Metropolitan Books, 2004.

Johnson, Janel. "Breaking the School-to-Prison Pipeline." *Civilrights.org,* 27 March 2008. http://www.civilrights.org/press_room/buzz_clips/civilrightsorg-stories/school-to-prison-pipeline.html (accessed 9 November 2008).

Johnson, Simon. "The Quiet Coup." *Atlantic Monthly,* 4 May 2009: 46–56.

Johnston, David Cay. "Corporate Wealth Share Rises for Top-Income Americans." *New York Times,* 29 January 2006: A1.

Jones, Dai, and Jonathan Elcock. *History and Theories of Psychology: A Critical Perspective.* London: Arnold, 2001.

Jones, Katherine. W. *Taming the Troublesome Child: American Families, Child Guidance, and the Limits of Psychiatric Authority.* Cambridge, MA: Harvard University Press, 1999.

Jordan, Miriam. "Report Warns of Influx of Hispanics in South Creates School Crisis." *Wall Street Journal,* 9 December 2004: B1-B2.

Joshi, Sopan. "Lack of Insurance, High Medical Costs Put More in a Bind." *Washington Post,* 20 August 2008: A2.

Jurik, Nancy. *Bootstrap Dreams: Microenterprise Development in an Era of Welfare Reform.* Ithaca, NY: ILR Press, 2005.

Kagan, Jerome. "Do Infants Think? *Scientific American* 226, no. 3 (March 1972): 74–82.

Kalita, Mitra S. "Americans See 18 Percent of Wealth Vanish." *Wall Street Journal,* 13 March 2009: A1, A8.

Kanner, Leo. *Child Psychiatry.* Springfield, IL: Charles C. Thomas, 1935.

Kantrowitz, Barbara, and Pat Wingert. "Unmarried with Children." *Newsweek,* 28 May 2001: 46.

Kaplan, Fred. "What's Really the U.S. Military Budget?" *Slate,* 4 February 2008. http://www.slate.com/toolbar.aspx?action=print$id=2183592 (accessed 7 January 2009).

Karmin, Craig. "Pension Bills to Surge Nationwide." *Wall Street Journal,* 16 March 2007: C1, C2.

Karnes, Merle, and Audrey Hodgins. "The Effects of a Highly Structured Preschool Program on the Measured Intelligence of Culturally Disadvantaged Four-Year-Old Children." *Psychology in the Schools* 6, no. 1 (1969): 89–91.

Karoly, Lynn A. "Rand Corporation: Forces Shaping the Future U.S. Workforce and Workplace Implications for 21st Century Work." Testimony presented before the House Education and Labor Committee on February 7, 2007. *Rand*

Corporation. http://www.rand.org/pubs/testimonies/CT273/ (accessed 9 March 2008).Karoly, Lynn A. and Constantijn W. Panis. "The 21st Century at Work: Forces Shaping the Future Workforce and Workplace in the United States." *Rand Corporation* 2004. http://www.rand.org/pubs/monographs/MG164/ (accessed 7 March 2008).

Karoly, Lynn, A. Bonnie Ghosh-Dastidar, Gail L. Zellman, Michal Perlman, and Lynda Fernyhough. "Prepared to Learn: The Nature and Quality of Early Care and Education for Preschool-Age Children in California." *Rand Corporation* 2008. http://www.rand.org/pubs/technical_reports/TR539/ (accessed 9 April 2008).

Katz, Michael B. *In the Shadow of the Poorhouse: A Social History of Social Welfare in America*. New York: Basic Books, 1996.

Kelley, Robert, E. "Managing the New Workforce." *Machine-Design* 62, no. 9 (1990): 109–13.

Kelly, Peter. "Youth at Risk: Process of Individualisation and Responsibilisation in the Risk Society." *Discourse: Studies in the Cultural Politics of Education* 22, no. 1 (2001): 23–33.

Kessen, William. "Early Settlements in New Cognition." *Cognition* 10 (1981): 167–71.

Key, Ellen. *The Century of the Child*. New York: Arno Press, 1972.

Keynes, John M. *The General Theory of Employment, Interest, and Money*. New York: Harcourt, 1965.

Kim-Cohen, J., A. Caspi, A. Taylor, B. Williams, R. Newcombe, I. W. Craig, and T. E Moffitt. "*MAOA*, Maltreatment, and Gene–Environment Interaction Predicting Children's Mental Health: New Evidence and a Meta-Analysis." *Molecular Psychiatry* 11 (2006): 903–13.

Kimmel, Michael. "Sociologist Michael Kimmel Comments on School Shootings." In *Major Problems in the History of American Families and Children: Documents and Essays*, edited by Anya Jabour, 500–501. Boston, MA: Houghton Mifflin, 2005.

Kiracofe, Clifford. "U.S. Imperialism: The National Security State." *Executive Intelligence Review*, 2 March 2006. http://www.larouchepub.com/other/2006/3311_berlin_kiracofe.html (accessed 3 March 2009).

Kirkpatrick, D. D. "Democrats in 2 Southern States Push Bills on Bible Study." *New York Times*, 27 January 2006. http://www.nytimes.com/2006/01/27/politics/27religion.html *(accessed* 27 January 2006).

Klare, Michael E. *Blood and Oil: The Dangers and Consequences of America's Growing Dependency on Imported Petroleum*. New York: Metropolitan Books, 2004.

Klein, Melanie. "Notes on Some Schizoid Mechanisms." In *Identity: A Reader*, edited by Paul DuGay, Jessica Evans, and Peter Redman, 130–43. London: Sage, 2000.

Kleinfield, N. R. "Diabetes and Its Awful Toll Quietly Emerge as a Crisis." *New York Times*, 9 January 2006: A1.

Kline, Wendy. *Building a Better Race: Gender, Sexuality, and Eugenics from the Turn of the Century to the Baby Boom*. Berkeley: University of California Press, 2001.

Kober, George. "The Progress and Achievements of Hygiene." *Science, New Series* 6, no. 152 (1897): 788–99.

Koch, Wendy. "Hunger Affecting More Low-Income Families. *USA Today*, 29 December 2008: 3A.

Kochhar, Rakesh. "Sharp Decline in Income for Non-Citizen Immigrant Households, 2006–2007." *Pew Hispanic Center,* 2 October 2008. http://pewhispanic.org/reports/report.php?ReportID=95 (accessed 9 November 2008).

Kolata, Gina B. "Childhood Hyperactivity: A New Look at Treatments and Causes." *Science, New Series* 199, no. 4328 (1978): 515–17.

Koprowski, Gene. "Bailout Cost Now Exceeds $7.7 Trillion." *MoneyNews,* 24 November 2008. http://moneynews.newsmax.com/streettalk/bailout_total_trillions/2008/11/24/154693.html (accessed 24 November 2008).

Koshuk, Ruth Pearson. "Social Influences Affecting the Behavior of Young Children." *Monographs of the Society for Research in Child Development* 6, no. (1941): i-iii, 1–71

Kozol, Jonathan. "Still Separate, Still Unequal: America's Educational Apartheid." *Harper's Magazine* 311, no. 1864 (2005). http://www.mindfully.org/Reform/2005/American-Apartheid-Education1sep05.htm (accessed 9 January 2009).

Lahart, Justin. "Jobless Rate Rises to 8.9% but Pace of Losses Eases." *Wall Street Journal,* 9–10 May 2009: A1–2.

Lahart, Justin, *Patrick Barta,* and *Andrew Batson.* "New Limits to Growth Revive Malthusian Fears; Spread of Prosperity Brings Supply Woes; Slaking China's Thirst." *Wall Street Journal,* 24 March 2008: A1.

Lakeoff, Andrew. "Adaptive Will: The Evolution of Attention Deficit Disorder." *Journal of the History of the Behavioral Sciences* 36, no. 2 (2000): 149–69.

Land, Kenneth C. The Child Well-Being Index (CWI) Report." *Foundation for Child Development,* May 2009. http://www.fcd-us.org/resources/resources_show.htm?doc_id=906348 (accessed 21 May 2009).

Landler, Mark, and David Jolly. "I.M.F. Puts Bank Losses from Global Financial Crisis at $4.1 Trillion." *New York Times,* 22 April 2009. http://www.nytimes.com/2009/04/22/business/global/22fund.html?th&emc=th (accessed 22 April 2009).

Langway, Lynn, Tenley-Ann Jackson, Marshal Zabarsky, Don Shirley, and James Whitmore. "Bringing Up Superbaby." *Newsweek,* 28 March 1983: 62

Lareau, Annette. *Unequal Childhood: Class, Race, and Family Life.* Berkeley: University of California Press, 2003.

Lasch, Christopher. *The Culture of Narcissism: American Life in an Age of Diminishing Expectations.* Rev. ed. New York: W. W. Norton, 1991.

Lasswell, Harold D. "The Garrison State." *American Journal of Sociology* 46, no. 3 (1941): 455–68.

Laufer, M. W., E. Denhoff, and G. Solomons. "Hyperkinetic Impulse Disorder in Children's Behavior Problems." *Psychomatic Medicine* 19 (1957): 38–49.

Lav, Iris J., and Elizabeth McNichol. "State Budget Troubles Worsen." *Center on Budget and Policy Priorities,* 13 March 2009. http://www.cbpp.org/cms/?fa=view&id=711 (accessed 19 April 2009).

Lavin, Timothy, and Jess Bachman. "Cash Machine." *Atlantic Monthly,* 4 May 2009: 58–59.

Layton, Lyndsey. "Probable Carcinogens Found in Baby Toiletries." *Washington Post,* 13 March 2009: A4.

Lemke, Thomas. "The Birth of Bio-Politics: Michel Foucault's Lecture at the Collège de France on Neo-Liberal Governmentality." *Economy and Society* 30 (2001): 190–207.

Levin, Ronnie, Mary Jean Brown, Michael E. Kashtock, David E. Jacobs, Elizabeth A. Whelan, Joanne Rodman, Michael R. Schock, Alma Padilla, and Thomas Sinks. "Lead Exposures in U.S. Children, 2008: Implications for Prevention." *Environmental Health Perspectives* 116 (May 2008):1285–93.

Levine, Mike. "Napolitano: As Recession Deepens, Illegal Immigration from Mexico Declines." *Fox News,* 9 February 2009. http://www.foxnews.com/politics/first100days/2009/02/09/napolitano-recession-deepens-illegal-immigration-mexico-declines/ (accessed 9 February 2009).

Levy, Clifford. "Emerging Powers Prepare to Meet in Russia." *New York Times,* 16 June 2009. http://www.nytimes.com/2009/06/16/world/europe/16bric.html?_r=1&th&emc=th (accessed 16 June 2009).

Lewis, Oscar. "The Culture of Poverty." *Scientific American* 215 (October 1966): 19–25.

Lindorff, Dave. "Bush's War on Children." *Counterpunch,* 24/25 May 2008. http://www.counterpunch.org/lindorff05242008.html (accessed 29 May 2008).

Lipman, Pauline. *High Stakes Education: Inequality, Globalization, and Urban School Reform.* New York: Routledge, 2003.

Lock, Margaret. "Biosociality and Susceptibility Genes: A Cautionary Tale." *Biosocialities, Genetics and the Social Sciences,* edited by Sahra Gibbon and Carlos Novas, 56–78. London: Routledge, 2008.

Locke, John. *Second Treatise of Government.* Ed. R. Cox. Arlington Heights, IL: Crofts Classics, 1982.

———. "Some Thoughts Concerning Education." In *Major Problems in the History of American Families and Children,* edited by A. Jabour, 75–77. New York: Houghton Mifflin Company, 2005.

Lohr, Steve. "Outsourcing Is Climbing Skills Ladder." *New York Times,* 16 February 2006. http://www.nytimes.com/ /02/16/business/16outsource.html (16 February 2006).

Lowenstein, Roger. "The End of Pensions." *New York Times,* 30 October 2005: A1.

Ludwig, Jens, and Susan Mayer. "'Culture' and the Intergenerational Transmission of Poverty: The Prevention Paradox." *Opportunity in America* 16, no. 2 (Fall 2006): 175–99.

Lueck, Sarah. "Action on Welfare Overhaul Is Stymied—State Officials Worry about How to Plan and Budget with Just a Short-Term Federal Fix." *Wall Street Journal,* 16 October 2002: A4.

Lumpe, Lora. "Small Arms Trade." *Foreign Policy in Focus* 3, no. 10 (1998). http://www.fpif.org/briefs/vol3/v3n10arms.html (accessed 9 March 2009).

Lydersen, Kari. "Boycott Underscores Disparities in Schools." *Washington Post,* 5 September 2008: A2.

Lyman, Rick. "In Many Public Schools, the Paddle Is No Relic." *New York Times,* 30 September 2006. http://www.nytimes.com/2006/09/30/education/30punish.html (accessed 30 September 2006).

Lyon, David. *Surveillance Society: Monitoring Everyday Life.* Buckingham: Open University Press, 2001.

Macfarquhar, Neil. "Donors' Aid to Poor Nations Declines, U.N. Reports." *New York Times,* 5 September 2008. http://www.nytimes.com/2008/09/05/world/05nations.html?_r=1&ref=world&oref=slogin (5 September 2008).

Macleod, D. I. *The Age of the Child: Children in America, 1890–1920.* New York: Twayne, 1998.

Macur, Juliet. "Born to Run? Little Ones Get Test for Sports Gene." *New York Times,* 30 November 2008: A1.

Maeroff, Gene I. "A Symbiosis of Sorts: School Violence and the Media." *Choices Briefs* 7 (2000): 3–9.

Mahtesian, Charles. "The Politics of Nature's Nurture." *Governing* 8, no. 11 (August 1995): 54.

Malacrida, Claudia. "Alternative Therapies and Attention Deficit Disorder." *Gender and Society* 16, no. 3 (2002): 366–85.

Marcus, Gary. *The Birth of the Mind: How a Tiny Number of Genes Creates the Complexities of Human Thought.* New York: Basic Books, 2003.

Marmot, Michael. *Status Syndrome: Your Social Standing Directly Affects Your Health and Life Expectancy.* London: Bloomsbury, 2004.

Marr, Kendra. "Children Targets of 1.6 Billion in Food Ads." *Washington Post,* 30 July 2008: D1.

Marshall, Haith M., and Joseph J. Campos. "Human Infancy." *Annual Review of Psychology* 28 (1977): 251–93.

Marshall, James D. "Foucault and Neo-liberalism: Biopower and Busno-Power." n.d.   http://www.ed.uiuc.edu/EPS/PES-Yearbook/95_docts/marshall.html (accessed 10 October 2004).

———. *Michel Foucault: Personal Autonomy and Education.* Dordrecht: Kluwer, 1996.

Martin, Randy. *An Empire of Indifference: American War and the Financial Logic of Risk Management.* Durham, NC: Duke University Press, 2007.

———. *The Financialization of Daily Life.* Philadelphia: Temple University Press, 2002.

Masland, Richard L., Seymour Sarason, and Thomas Gladwin. *Mental Subnormality: Biological, Psychological and Cultural Factors.* New York: Basic Books, 1958.

Maslow, A. H. *Toward a Psychology of Being* 2nd ed. New York: D. Van Nostrand Company, 1968.

"Mass Killings in Eastern Congo." *New York Times,* 11 December 2008. http://video.nytimes.com/video/2008/12/11/world/africa/1194835234624/mass-killings-in-eastern-congo.html?th&emc=th (accessed 11 December 2008).

Matthews, Hannah. "Child Care Assistance Helps Families Work: A Review of the Effects of Subsidy Receipt on Employment." Center for Law and Social Policy (CLASP), 3 April 2006. http://www.clasp.org/publications/ccassistance_employment.pdf (accessed 9 May 2008).

Matthews, Robert G. "Median Household Income Rises 1.1%." *Wall Street Journal,* 30 August 2006: A2.

Mayes, Rick, and Adam Rafalovich. "Suffer the Restless Children: The Evolution of ADHD and Paediatric Stimulant Use, 1900–80." *History of Psychiatry* 18, no. 4 (2007): 435–57.

Macfarquhar, Neil. "Donors' Aid to Poor Nations Declines, U.N. Reports." *New York Times,* 5 September 2008. http://www.nytimes.com/2008/09/05/world/05nations.html?_r=1&ref=world&oref=slogin (accessed 5 September 2008).

McCallum, D. *Personality and Dangerousness: Genealogies of Antisocial Personality Disorder.* Cambridge: Cambridge University Press, 2001.

McHugh, David. "U.N. Survey Ranks Britain, U.S. Low in Child Welfare." *Arizona Republic,* 15 February 2007: A17.

McCrummen, Stephanie. "Report: U.S. Africa Aid Is Increasingly Military; Advocacy Group Cites Development Needs." *Washington Post*, 18 July 2008: A10.

McGough, Robert. "More Kids Given Antidepressants as Number of Diagnoses Rockets." *Wall Street Journal*, 5 May 2004: D2.

Meckel, R. A. *Save the Babies: American Public Health Reform and the Prevention of Infant Mortality 1850–1929*. Baltimore: Johns Hopkins University Press, 1990.

Meckler, Laura. "How a U.S. Official Promotes Marriage to Help Poor Kids." *Wall Street Journal*, 20 November 2006: A1.

Medley, Sara Sullivan. "Childhood Lead Toxicity: A Paradox of Modern Technology." *Annals of the American Academy of Political and Social Science* 461 (1982): 63–73.

"Medvedev Calls for Use of National Currencies in Trade." *Voice of Russia*, 16 June 2009. http://ruvr.ru/main.php?lng=eng&q=46798&cid=215&p=16.06.2009 (accessed 16 June 2009).

Merle, Renae. "Defense Earnings Continue to Soar." *Washington Post*, 30 July 2007: D1.

———. "Recalls of Toys Pressure Agency." *Washington Post*, 3 August 2007: D1.

Merlo, Due P., Y. Harel-Fisch, M. T. Damsgaard, B. E. Holstein, J. Hetland, C. Currie, S. N. Gabhainn, M. G. deMatos, and J. Lynch. "Socioeconomic Inequality in Exposure to Bullying during Adolescence: A Comparative, Cross-Sectional, Multilevel Study in 35 Countries." *American Journal of Public Health* 99, no. 5 (2009): 907–14.

Meyer, Pamela, Timothy Pivetz, Timothy A. Dignam, David M. Homa, Jaime Schoonover, and Debra Brody. "Surveillance for Elevated Blood Levels among Children—United States, 1997–2001." *Surveillance Summaries* 52, no. SS10 (September 2003): 1–21.

Milken, Michael. "Amid Plenty, the Wage Gap Widens." *Wall Street Journal*, 5 September 2000. http://webreprints.djreprints.com/0000000000000000018456001.html (accessed 9 May 2009).

Miller, Paul, Charles Mulvey, and Nick Martin. "Genetic and Environmental Contributions to Educational Attainment in Australia." *Economics of Education Review* 20, no. 3 (2001): 211–24.

Miller, Toby. *Makeover Nation: The United States of Reinvention*. Columbus: Ohio State University Press, 2008.

Miller, Toby, and Marie Claire Leger. "A Very Childish Moral Panic: Ritalin." *Journal of Medical Humanities* 24, nos. 1/2 (2003): 9–33.

Millichap, Gordon, J. F. Aymat, L. H. Sturgis, K. W. Larsen, and R. A. Egan. "Hyperkinetic Behavior and Learning Disorders: III. Battery of Neuropsychological Tests in Controlled Trial of Methylphenidate." *American Journal of Diseases of Children* 116, no. 3 (1968): 235–44.

Mintz, Steven. *Huck's Raft: A History of American Childhood*. Cambridge, MA: The Belknap Press, 2006.

Mishkin, Frederic. "On 'Leveraged Losses: Lessons from the Mortgage Meltdown' at the U.S. Monetary Policy Forum, New York, New York." Board of Governors of the Federal Reserve System, 29 February 2008: http://www.federalreserve.gov/newsevents/speech/mishkin20080229a.htm (accessed 9 March 2009).

Monahan, Torin, ed. *Surveillance and Security: Technological Politics and Power in Everyday Life*. New York: Routledge, 2006.

Monbiot, George. "Behind the Phosphorus Clouds are War Crimes within War Crimes." *Guardian,* 22 November 2005. http://www.guardian.co.uk/world/2005/nov/22/usa.iraq1 (9 December 2006).

Montagu, Ashley. "Sociogenic Brain Damage." *American Anthropologist, New Series* 74, no. 5 (October 1972): 1045–61.

Moore, Solomon. "Missouri System Treats Juvenile Offenders with Lighter Hand." *New York Times,* 27 March 2009. http://www.nytimes.com/2009/03/27/us/27juvenile.html?_r-&th=&emc-th&pagewanted (accessed 27 March 2009).

Moore, Solomon. "Troubles Mount within Texas Youth Detention Agency. *New York Times,* 16 October 2007: A1.

Moore, Solomon. "Gangs Grow, but Hard Line Stirs Doubts. *New York Times,* 13 September 2007. http://www.nytimes.com/2007/09/13/us/13gang.html (accessed 13 September 2007).

Moravcik, Meghan, E. "Helping Kids Balance the Facts: Starting a Financial Education Early is Key to Raising Money-Wise Children." *Arizona Republic,* 14 September 2008: B4.

Morris, C., A. Shen, K. Peirce, and J. Beckwith. "Deconstructing Violence." *GeneWatch* 20, no. 2 (2007). http://www.gene-watch.org/genewatch/articles/20–2Beckwith.html (accessed 9 May 2008).

Morrison, Blake, and Brad Heath. "Health Risks Stack Up for Students Near Industrial Plants. *USA Today,* 7 December 2008. http://www.usatoday.com/news/nation/environment/school-air1.htm (accessed 7 December 2008).

Morrison, J. R., and M. A. Stewart. "A Family Study of the Hyperactive Child Syndrome." *Biological Psychiatry* 3 (1971): 189–95.

Morss, John R. *The Biologising of Childhood: Developmental Psychology and the Darwinian Myth.* Hove: Lawrence Erlbaum, 1990.

Moss, William. "An Essay on the Management, Nursing and Diseases of Children, from the Birth: And on the treatment and Diseases of Pregnant and Lying-In Women, which was Defigned for Domeftic Ufe , and Purpfely Adapted for Female Comprehension." Philadelphia: Thomas Samuel and T. Bradford, 1794. *Eighteenth Century Collections Online.* Gale Group. http://callisto.ggsrv.com/imgsrv/Fetch?banner=441f324e&digest=b2897d49932809afde18 (9 May 2007).

Moynihan, Daniel Patrick. "The Negro Family: The Case For National Action." Office of Policy Planning and Research United States Department of Labor, March 1965. http://www.dol.gov/oasam/programs/history/webid-meynihan.htm (accessed 9 April 2008).

Munk, N. "New Organization Man." *Fortune,* 16 March 1998: 63–74.

Murphy, Elizabeth. "Expertise and Forms of Knowledge in the Government of Families." *Sociological Review* 51 no. 4 (2003): 433–62.

Murray, Charles. "On Education: Intelligence in the Classroom." *Wall Street Journal,* 16 January 2007: A20.

Murray, Sara. "The Curse of the Class of 2009." *Wall Street Journal,* 9–10 May 2009: A1, A11.

Nadesan, Majia. *Constructing Autism: Unravelling the 'Truth' and Discovering the Social.* London: Routledge, 2005.

———. "Engineering the Entrepreneurial Infant: Brain Science, Infant Development Toys, and Governmentality." *Cultural Studies* 16, no. 3 (2002): 401–32.

———. *Governmentality, Biopower, and Everyday Life.* New York: Routledge, 2008.

Nadesan, Majia. "Hurricane Katrina: Governmentality, Risk, and Responsibility." *Controversia* 5, no. 2 (2008): 67–90.

———. "The Make Your Day Panopticon: Neoliberalism, Governmentality and Education." *Radical Pedagogy* 8, no. 1 (2006): http://radicalpedagogy.icaap.org/currentissue.html.

———. "The Popular Success Literature and 'A Brave New Darwinian Workplace.'" *Consumption, Markets & Culture* 3, no. 1 (1999): 27–60.

Nadesan, Majia, and Patty Sotirin. "The Romance and Science of 'Breast is Best': Discursive Contradictions and Contexts of Breast-feeding Choices." *Text and Performance Quarterly* 18 (1998): 217–32.

Nagourney, E. "Handing Baby a Start on Genius." *Arizona Republic*, 29 November 2002: A12.

Naik, Gautam. "Long Division: The Debate over the Value of Preschool." *Wall Street Journal*, 29 August 2008. http://online.wsj.com/article/SB121997547720682181.html (accessed 29 August 2008).

National Center for Public Policy Research. "The Social Security Act of 1935." The National Center, Social Security Administration. n.d. http://www.nationalcenter.org/SocialSecurityAct.html (accessed 7 May 2008).

National Institutes of Health. "Bullying Widespread in U.S. Schools, Survey Finds." 21 April 2001. http://www.nichd.nih.gov/news/releases/bullying.cfm (accessed 29 May 2008).

National Resource Defense Council. "The Story of *Silent Spring*." National Resource Defense Council, 16 April 1998. http://www.nrdc.org/health/pesticides/hcarson.asp (accessed 12 December 2008).

Nazario, Sonia L. "Education: Schools Teach the Virtues of Virginity." *Wall Street Journal*, 20 February 1992: B1.

Needleman, Herbert. L. "Lead Poisoning in Children: Neurological Implications of Widespread Subclinical Intoxication." *Seminars in Psychiatry* 5 (1973): 47–54.

Needleman, H., J. Riess, M. Tobin, G. E. Biesecker, and J. B. I Greenhouse. "Bone Lead Levels and Delinquent Behavior." *Journal of the American Medical Association* 275, no. 5 (1996): 363–69.

Needleman, Herbert, Charles Gunnuoe, Alan Leviton, Robert Reed, Henry Peresie, Cornelius Maher and Peter Barrett. "Deficits in Psychologic and Classroom Performance of Children with Elevated Dentine Lead Levels," *New England Journal of Medicine* 300, no. 13 (March 1979): 689–93.

Nelkin, Dorothy. "Behavioral Genetics and Dismantling the Welfare State." In *Behavioral Genetics: The Clash of Culture and Biology*, edited by Ronald A. Carson and Mark A. Rothstein, 156–71. Baltimore: Johns Hopkins University Press, 1999.

———. "Molecular Metaphors: The Gene in Popular Discourse." *Nature* 2, no. 7 (2001): 556–59.

Nelson, Harry. "Possible Link of Lead in Blood, Smog Studied." *Los Angeles Times*, 6 June 1961: B3.

———. "Smog Facts: Air Pollution May Stay as Problem, Say Experts." *Los Angeles Times*, 8 January 1961: I1.

Nelson, Margaret, and Rebecca Schutz. "Day Care Differences and the Reproduction of Social Class." Paper presented at the annual meeting of the American Sociological Association, Montreal Convention Center, Montreal, Quebec, Canada, 10 August 2006. http://www.allacademic.com/meta/p95387_index.html (accessed 9 May 2008).

Nelson-Rowe, S. "Ritual, Magic, and Educational Toys: Symbolic Aspects of Toy Selection." In *Troubling Children*, edited by Joel Best, 117–31. Hawthorne, NY: Aldine de Gruyter, 1994.

Nettleton, Sarah. "Wisdom, Diligence and Teeth: Discursive Practices and the Creation of Mothers." *Sociology of Health and Illness* 13, no.1 (1991): 98–111.

Nevin, Rick. "Understanding International Crime Trends: The Legacy of Preschool Lead Exposure. *Environmental Research* 104 (2007): 315–36.

"New Freedom Commission on Mental Health Report." Mental Health Commission. 29 April 2002. http://www.mentalhealthcommission.gov/reports/FinalReport/downloads/FinalReport.pdf (accessed 9 June 2008).

Nichiporuk, Brian. *Security Dynamics of Demographic Factors*. Santa Monica: Rand Corporation, 2007. http://www.rand.org/pubs/monograph_reports/MR1088/ (accessed 8 June 2008).

Nietzsche, Friedrich. *The Genealogy of Morals*. Ed. Walter Kaufmann. New York: Vintage, 1989.

Nimr, Heba, Catherine Tactaquin, and Arnoldo Garcia. "Human Rights and Human Security at Risk: The Consequences of Placing Immigration Enforcement in the Department of Homeland Security." U.S. National Network for Immigrant and Refugee Rights, 2003. http://www.nnirr.org/get/get_dhs.html (accessed 10 April 2009).

Nisbett, Richard E. "All Brains Are the Same Color." *New York Times,* 9 December 2007. http://www.nytimes.com/2007/12/09/opinion/09nisbett.html *(accessed* 9 December 2007).

Nordstrom, Carolyn. *Shadows of War: Violence, Power, and International Profiteering in the Twenty-First Century*. Berkeley: University of California Press, 2004.

Nutting, M. Adelaide. "The Home and Its Relation to the Prevention of Disease." *The American Journal of Nursing* 4, no. 12 (September 1904): 913–24.

Nutting, Rex. "U.S. Corporate Profits Fell in the Fourth Quarter of 2006, Signaling the End of One of the Greatest Profit Cycles in Post-War Era, Economists Say." *MarketWatch,* 29 March 2007. http://www.marketwatch.com/story/us-corporate-profits-have-peaked-economists-say (5 May 2009).

Olssen, Mark. *Michel Foucault: Materialism and Education*. Westport, CT: Bergin & Garvey, 1999.

O'Malley, Pat. *Risk, Uncertainty and Government*. London: GlassHouse, 2004.

O'Malley, Pat, and Steven Hutchinson. "Reinventing Prevention: Why Did 'Crime Prevention' Develop So Late? *British Journal of Criminology* 47 (2006): 373–89.

Ohadi, Mina, Elham Shirazi, Tehranidoosti Mehdi, Narges Moghimi, Mohammad Keikhaee, Sima Ehssani, Ali Aghajani, and Hossein Najmabadi. "Attention-Deficit/Hyperactivity Disorder (ADHD) Association with the DAT1 Core Promoter-67 T Allele." *Brain Research* 1101, no. 1 (2006): 1–4.

Ohlemacher Stephen. "Persistent Race Disparities Found." *Washington Post,* 14 November 2006: A3.

Olim, Ellis. "Maternal Language Styles and Children's Cognitive Behavior." *Journal of Special Education* 4, no. 1 (1970): 53–68.

Olim, E., R. Hess, and V. Shipman. "Role of Mothers' Language Styles in Mediating Their Pre-School Children's Cognitive Development." *School Review* 75, no. 4 (1967): 414–24.

Osmundsen, John A. "Scientist Calls for a New Study of Negro-White 'Differences.' " *New York Times,* 18 October 1964: 73.

Ossorio, P., and T. Duster, T. "Race and Genetics: Controversies in Biomedical, Behavioral, and Forensic Sciences." *American Psychologist* 60 (2005): 115–28.

"Paint Eaters." *Time* archives, 20 December 1943. http://www.time.com/time/magazine/artlce/0,9171,932645,00.html (accessed 8 June 2008).

Palermo, Joseph. "Socialism for Wall Street Capitalism for Main Street." *The Huffington Post*, 17 September 2008. http://www.huffingtonpost.com/joseph-a-palermo/socialism-for-the-rich-na_b_127121.html (accessed 8 March 2009).

Paley, Amit R. "A Quiet Windfall for U.S. Banks." *Washington Post*, 10 November 2008: A1.

Parker-Pope, Tara. "A New Face for ADHD and a Debate." *New York Times*, 25 November 2008. http://www.nytimes.com/2008/11/25/health/25well.html?em=&pagewanted=print (accessed 25 November 2008).

Parrott, Sharon. "Recession Could Cause Large Increases in Poverty and Push Millions into Deep Poverty." Center for Budget and Policy Priorities, 24 November 2008. http://www.cbpp.org/cms/index.cfm?fa=view&id=2228 (accessed 20 June 2009).

Partridge, T. "Are Genetically Informed Designs Genetically Informative? Comment on McGue, Elkins, Walden, and Iacono (2005) and Quantitative Behavioral Genetics." *Developmental Psychology* 41, no. 6 (2005): 985–88.

Patterson, Orlando. "The Jailing of Black America." *New York Times*, 2 October 2007. http://www.nytimes.com/2007/10/02/opinion/02iht-edpatterson.1.7716628.html (accessed 2 October 2007).

Payne, Ruby K. "A Framework for Understanding Poverty." 2005. http://www.ahaprocess.com/About_Us/Ruby_Payne.html (accessed 9 May 2008).

Pearlstein, Steven. "Keynes on Steroids." *New York Times*, 26 November 2008: D1.

Pereira, Joseph. "Protests Spur Stores to Seek Substitute for Vinyl in Toys. *Wall Street Journal*, 12 February 2008: B1.

Peters, Michael. *Education, Globalization, and the State in the Age of Terrorism.* Boulder, CO: Paradigm, 2005.

Petersen, Andrea. "To Be Young and Anxiety Free." *Wall Street Journal*, 2 September 2008: D1-D2.

Petersen, Alan, and Deborah Lupton. *The New Public Health: Health and Self in the Age of Risk.* London: Sage, 1996.

Peterson, Rita. "Great Expectations: Collaboration between the Brain Sciences and Education." *American Biology Teacher* 46, no. 2 (1984): 74–80.

Petrecca, Laura. "Financially Pinched Companies Snip Employee Benefits." *USA Today*, 7 April 2009. http://www.usatoday.com/money/workplace/2009–04-06-employers-cut-worker-benefits_N.htm (accessed 15 June 2009).

Phillips, Kevin. "The Disaster Stage of U.S. Financialization." *TPM Café*, 7 April 2009. http://tpmcafe.talkingpointsmemo.com/2009/04/07/the_disaster_stage_of_us_financialization/ (accessed 8 May 2009).

Piaget, Jean. *Judgment and Reasoning in the Child.* Trans. M. Warden. Savage, MD: Littlefield, Adams and Co., 1972.

———. *The Language and Thought of the Child.* Trans. M. Gabain. London: Routledge and Kegan Paul, 1959.

Pilkington, Ed. "Jailed for a MySpace Parody, the Student Who Exposed America's Cash for Kids Scandal." *Guardian*, 7 March 2009. http://www.guardian.co.uk/world/2009/mar/07/juvenille-judges-cash-detention-centre (accessed 8 April 2009).

Pincus, Walter. "Climate Issues Tied to U.S. Security." *Washington Post,* 26 June 2008: A2

Platt, Anthony M. *The Child Savers: The Invention of Delinquency.* Chicago: University of Chicago Press, 1969.

Plomin, Robert, and Ian Craig. "Epidemiology in Neurobiological Research: Genes, Environments and Cognitive Abilities." *British Journal of Psychiatry* 178 (2001): s41-s48.

Pogash, Carol. "Free Lunch Isn't Cool, So Some Students Go Hungry." *New York Times, 1 March 2008.* http://www.nytimes.com/2008/03/01/education/01lunch.html (accessed *1 March 2008*).

"Police Immigrant Killed by Teens Because of Race." *NBC New York News,* 11 November 2008. http://www.nbcnewyork.com/news/local/Police-Immigrant-Killed-Because-Of-Race-.html (accessed 8 January 2009).

Pollock, L. A. Foreword. *Picturing Children: Constructions of Childhood between Rousseau and Freud.* Ed. M. R. Brown. Aldershot, UK: Ashgate, 1992.

Popkewitz, Thomas. S. *Struggling for the Soul: The Politics of Schooling and the Construction of the Teacher.* New York: Teachers College Press, 1998.

Preston, S. H., and Michael R. Haines. *Fatal Years: Child Mortality in Late Nineteenth Century America.* Princeton, NJ: Princeton University Press, 1991.

Proctor, Robert. "Genomics and Eugenics: How Fair Is the Comparison?" In *Gene Mapping: Using Law and Ethics as Guides,* edited by George J. Annas and Sharman Elias, 57–93. New York: Oxford University Press, 1992.

"Programmed Obesity." *Environmental Health Perspectives* 117, no. 1 (January 2009). http://www.ehponline.org/docs/2009/117-1/ss.html#game (accessed 20 January 2009).

Quindlen, Anna. "The Good Enough Mother." *Newsweek,* 21 February 2005: 50.

Rabinow, Paul. "Artificiality and Enlightenment: From Sociobiology to Biosociality." In *Anthropologies of Modernity: Foucault, Governmentality, and Life Politics,* edited by J. X. Inda, 181–93. Madden, MA: Blackwell, 2005.

Kochhar, Rakesh. "Sharp Decline in Income for Non-Citizen Immigrant Households, 2006–2007." *Pew Hispanic Center,* 2 October 2008. http://pewhispanic.org/reports/report.php?ReportID=95 (accessed 11 December 2008).

Ramesh, Randeep. "Children Still Dying in Booming Asian Economies." *Guardian,* 5 August 2008. http://www.guardian.co.uk/world/2008/aug/05/india.china (accessed 5 August 2008).

Reagan, Michael, B. "US Military Recruits Children: 'America's Army' Video Game Violates International Law." *Truthout,* 23 July 2008. http://www.truthout.org/article/us-military-recruits-children (accessed 23 July 2008).

"Recession, Bailout Costs Push Deficit to Record." *MSNBC,* 11 February 2009. http://www.msnbc.msn.com/id/29144295/ (accessed 11 February 2009).

Reed, Jane. "Lead Poisoning: Silent Epidemic and Social Crime." *American Journal of Nursing* 72, no. 12 (1972): 2180–84.

Reef, Catherine. *Childhood in America.* New York: Facts on File, 2002.

———. *Poverty in America.* New York: Facts on File, 2007.

———. *Working in America.* New York: Facts on File, 2000.

Reich, Robert. "When Will the Recovery Begin? Never." *Robert Reich's Blog,* 9 July 2009. http://robertreich.blogspot.com/2009/07/when-will-recovery-begin-never.html (accessed 11 July 2009).

Reichman, N. "Managing Crime Risks: Toward an Insurance Based Model of Social Control." *Research in Law and Social Control* 8 (1986):151–72.

Reinhold, Robert. "Rx for Child's Learning Malady." *New York Times*, 3 July 1970: 27.

Rensberger, Boyce, and Philip J. Hilts. "The Role of Genetics in IQ Scores." *Washington Post*, 9 January 1989: A3.

"Researchers Learning How Young Brains Grow." *Arizona Republic*, 19 May 2007: A19.

Reznikoff, Paul. "Lead Poisoning." *American Journal of Nursing* 42, no. 10 (1942): 1123–26.

Richardson, Teresa. *The Century of the Child: The Mental Hygiene Movement and Social Policy in the United States and Canada*, Albany: State University of New York Press, 1989.

Riggs, Marlon. *Ethnic Notions Marlon Riggs*. Producer/Director Marlon Riggs. Narrator Esther Rolle. California Newsreel, 1987.

Roa, Brian. "The Chicago Model of Militarizing Schools." *Truthout*, 29 June 2009. http://www.truthout.org/062909T (accessed 2 July 2009).

Roberts, Sam. "Minorities Often a Majority of the Population under 20." *New York Times*, 7 August 2008: A1

Rogers, Walter. "America: A Superpower No More: Decline is Occurring More Rapidly Than We Think. It's Time to Embrace a New Agenda" *Christian Science Monitor*, 8 April 2009. www.csmonitor.com/2009/0408/p09s01-coop.html (accessed 8 April 2009).

Rose, Nikolas. *Governing the Soul: The Shaping of the Private Self.* London: Routledge, 1990.

———. *The Politics of Life Itself.* Princeton, NJ: Princeton University Press, 2007.

———. *Powers of Freedom.* Cambridge: Cambridge University Press, 1999.

Rosenfeld, Alvin. *The Overscheduled Child.* New York: St. Martin's, 2001.

Rosner, D., and Markowitz, G. "A 'Gift of God'? The Public Health Controversy over Leaded Gasoline during the 1920s." *American Journal of Public Health* 75 (1985): 344–52.

Rothschild, Joan. *The Dream of the Perfect Child.* Bloomington, IN: Indiana University, 2005.

Roubini, Nouriel. "Worst Economic and Financial Crisis since the Great Depression Reveals the Weaknesses of the Laissez-Faire Anglo-Saxon Model of Capitalism." *RGE Monitor*, 19 February 2009. http:www.rgmonitor.com/roubini-monitor/255627/the_worst_economic_and_financial_crisis_since_the_great_depression_reveals_the_weaknesses_of_the_laissez_faire_anglo-saxon_model_of_capitalism (accessed 19 February 2009).

———. "U.S. Government Finance." *RGE Monitor*, 11 March 2009. http://www.rgemonitor.com/662?cluster_id=9631 (accessed 11 March 2009).

Rubin, A. J. "In Seized Video, Boys Train to Fight in Iraq, U.S. Says." *New York Times*, 7 February 2008. http://www.nytimes.com/2008/02/07/world/middleeast/07iraq.html (accessed 7 February 2008).

Rubins, Nancy . "The Truth about Hyperactivity." *Parents Magazine*, 64 (February 1985): 110–12.

Rush, Benjamin *Essays: Literary, Moral, and Philosophical.* Philadelphia: Thomas S. Samuel & F. Bradford, 1798.

Rutherford, J. "At War." *Cultural Studies* 19 (2005): 622–42.

"SCO, BRIC Urge Fairer World Order." *Voice of Russia*, 17 June 2009. http://ruvr.ru/main.php?lng=eng&q=46845&cid=206&p=17.06.2009 (accessed 18 June 2009).

Sampson, Rana. "Bullying in Schools. Problem Oriented Guides for Police Series." U.S. Department of Justice Office of Community-Related Policing Services, 2002. http://www.cops.usdoj.gov/pdf/e12011405.pdf (accessed 9 April 2008).

Sard, Barbara. "Number of Homeless Families Climbing Due to Recession." Center on Budget and Policy Priorities, 8 January 2009. http://www.cbpp.org/cms/index.cfm?fa=view&id=2228 (accessed 20 June 2009).

Saudino, Kimberly J., and Robert Plomin. "Why Are Hyperactivity and Academic Achievement Related?" *Child Development* 78, no.3 (2007): 972–86.

Sapien, Joaquin, and ProPublica. "CDC Study Finds Rocket Fuel Chemical in Baby Formula." *Scientific American,* 9 April 2009. http://www.scientificamerican.com/article.cfm?id=cdc-study-finds-rocket-fuel-chemical-in (accessed 9 April 2009).

Schlesinger, Arthur. *The Imperial Presidency.* Boston, MA: Houghton Mifflin, 1973.

Schneider, Howard. "Carter Decries Gaza Curbs, Asks Israel to Halt 'Abuse.'" *Washington Post,* 17 June 2009: A1.

Schrag, Peter, and Diane Divoky. *The Myth of the Hyperactive Child: And Other Means of Child Control.* New York: Pantheon, 1975.

Schram, Sanford F. "Neoliberal Poverty Governance: U.S. Welfare Policy in an Era of Globalization." Presentation at "The New Poverty Agenda: Reshaping Policies in the 21st Century." Queen's University, International Institute on Social Policy Kingston, Ontario, 18–20 August 2008. http://www.queensu.ca/sps/conferences_events/qiisp/2008/Schram.pdf (accessed 7 November 2008).

Schwarz, Conrad J. "Childhood Origins of Psychopathology." *American Psychologist* 34, no. 10 (1979): 879–85.

Schweid, Barry. "Study: US Arms Sales Undermine Global Human Rights." *San Francisco Chronicle,* 10 December 2008. http://www.sfgate.com/cgi-bin/article.cgi?f=/n/a/2008/12/10/national/w104346S60.DTL (accessed 9 January 2009).

Scott, Janny. "Life at the Top in America Isn't Just Better, It's Longer. *New York Times,* 16 May 2005. http://www.nytimes.com/2005/05/16/national/class/HEALTH-FINAL.html (accessed 16 May 2005).

Sealander, Judith. *The Failed Century of the Child: Governing America's Young in the Twentieth Century.* Cambridge: Cambridge University Press, 2003.

Sedgwick, William T., and Theodore Hough. "What Training in Physiology and Hygiene May We Reasonably Expect of the Public Schools?" *Elementary School Teacher* 4, no. 3 (1903): 132–44.

Seiter, Ellen. *Sold Separately: Parents and Children in Consumer Culture.* New Brunswick, NJ: Rutgers University Press, 1993.

Selling, L. S. "Restlessness in a Delinquent Group." *Psychological Clinic* 20 (1931): 92–93.

Semple, Kirk. "A Killing in a Town Where Latinos Sense Hate." *New York Times,* 14 November 2008: A25.

Shanker, Thom. "Command for Africa Is Established by Pentagon." *New York Times,* 5 October 2008. http://www.nytimes.com/2008/10/05/world/africa/05command.html (accessed 5 October 2008).

———. "U.S. Is Top Arms Seller to Developing World." *New York Times,* 1 October 2007. http://query.nytimes.com/gst/fullpage.html?res=9D01E0DA1739F932

A35753C1A9619C8B63&sec=&spon=&pagewanted=all (accessed 1 October 2007).

———. "Weapons Sales Worldwide Rise to Highest Level since 2000." *New York Times,* 30 August 2005. http://www.nytimes.com/2005/08/30/politics/30weapons.html (accessed 30 August 2005).

Shapira, Ian. Author's Poverty Views Disputed yet Utilized." *Washington Post,* 15 April 2007: A1.

Shin, Annys. "A Coach at the Crib and a Consultant at the Potty." *Washington Post,* 8 March 2008: A1.

———. "Taking Lead Safety into Its Own Hands." *Washington Post,* 11 November 2007: D1.

———. "Tighter Lead Rule for Kids' Items." *Washington Post,* 15 September 2009: A9.

Shin, Annys, and Neil Irwin. "Economy Shrinks at Staggering Rate." *Washington Post,* 28 February 2009: A1.

Shipler, David K. *The Working Poor: Invisible in America.* New York: Vintage, 2005.

Sibley, David. *Geographies of Exclusion: Society and Difference in the West.* New York: Routledge, 1995.

Singer, E. *Child-Care and the Psychology of Development.* Trans. A. Porcelijn. London: Routledge, 1992.

Singer, Peter W. *Children at War.* New York: Pantheon Books, 2004.

———. "Young Soldiers Used in Conflicts around the World." *Brookings Institute,* 12 June 2006. http://www.brookings.edu/interviews/2006/0612humanrights_singer.aspx (accessed 9 November 2008).

Singh, Ilina. "Doing Their Jobs: Mothering with Ritalin in a Culture of Mother-Blame." *Social Science and Medicine* 59 (2004): 1193–1205.

Slackman, Michael. "In Algeria, a Tug of War for Young Minds." *New York Times,* 23 June 2008. *http://www.nytimes.com/2008/06/23/world/africa/23algeria.html (accessed 23 June 2008).*

Smith, Adam. *An Inquiry into the Nature and Causes of the Wealth of Nations.* 3rd ed. Edited by R. H. Campbell and A. S. Skinner. Oxford: Clarendon Press, 1976. Original work published 1784.

Smith, Jonathan. Z. "Religion, Religions, Religious." In *Critical Terms for Religious Studies,* edited by M. C. Taylor, 269–84. Chicago: University of Chicago Press, 1998.

Smith, Stephen. "The Practice of Public Health, 1952: A Special Section: Part II." *Public Health Reports* 68, no. 2 (February 1953): 197–280.

Sontag, Marvin, Adina Sella, and Robert L. Thorndike. "The Effect of Head Start Training on the Cognitive Growth of Disadvantaged Children." *Journal of Educational Research* 62, no. 9 (1969): 387–89.

Spaulding, Edith R., and William Healy. "Inheritance as a Factor in Criminality." *Journal of the American Institute of Criminal Law and Criminology* 4 (1914): 837–58.

Speak, Suzanne, and Graham Tipple. "Perceptions, Persecution and Pity: The Limitations of Interventions for Homeless in Developing Countries." *International Journal of Urban and Regional Research* 30 (2006): 172–88.

Spencer, Naomi. "US: 12 Million Children Face Hunger and Food Insecurity." World Socialist Web Site, 11 May 2009. http://www.wsws.org/articles/2009/may2009/food-m11.shtml (accessed 15 June 2009).

Spring, Joel. *Pedagogies of Globalization: The Rise of the Educational Security State.* Mahwah, NJ: LEA, 2006.

Springer, John L. "Small Wonder Called the Gene." *New York Times,* 23 November 1958: 15, 24–25.

St. George, Donna. "Study Finds Kids' Activities Linked to Success, Not Stress." *Arizona Republic,* 28 September 2008: A21.

———. "Study Finds Some Youths 'Addicted' to Video Games." *Washington Post,* 20 April 2009: A2.

Stiglitz, Joseph. "Wall Street's Toxic Message." *Vanity Fair,* July 2009. http://www.vanityfair.com/politics/features/2009/07/third-world-debt200907 (accessed 9 July 2009).

Stearns, Peter N. *Anxious Parents: A History of Modern Childrearing in America.* New York: New York University Press, 2003.

Stein, Rob. "Fetal Gene Testing Spurs Hope, Alarm." *Arizona Republic,* 26 October 2008: A8.

———. "Research Links Poor Kids' Stress, Brain Impairment." *Washington Post,* 6 April 2009: A6.

Stephens, Sharon. "Children and the Politics of Culture in 'Late Capitalism.'" In *Children and the Politics of Culture,* edited by Sharon Stephens, 3–50. Princeton, NJ: Princeton University Press, 1995.

Stern, Alexandra Minna. *Eugenic Nation: Faults and Frontiers of Better Breeding in Modern America.* Berkeley: University of California Press, 2005.

Stevenson, G. S. *Child Guidance Clinics: A Quarter Century of Development.* New York: The Commonwealth Fund, Oxford University Press, 1934.

Stewart, M., and J. Morrison. "Affective Disorder among the Relatives of Hyperactive Children." *Journal of Child Psychology and Psychiatry* 14 (1973): 209–12.

Stiglitz, Joseph E. "The Broken Promise of Nafta." *New York Times,* 6 January 2004. http://www.nytimes.com/2004/.../the-broken-promise-of-nafta.html (accessed 6 January 2004).

———. "Wall Street's Toxic Message." *Vanity Fair,* 29 July 2009.

Still, George F. "Some Abnormal Psychical Conditions in Children. Lecture I." *The Lancet,* 12 April 1902: 1008–13.

———. "Some Abnormal Psychical Conditions in Children. Lecture II." *The Lancet,* 19 April 1902: 1077–82.

———. "Some Abnormal Psychical Conditions in Children. Lecture III." *The Lancet,* 16 April 1902: 1163–68.

Stone, M. H. "Child Psychiatry before the Twentieth Century." *International Journal of Child Psychotherapy* 2 (1973): 264–308.

Story, L. "Citizen Vigilance Leads to Toy Recalls." *New York Times,* 22 November 2007. http://www.nytimes.com/2007/11/22/business/22recall.html (accessed 22 November 2007).

Straziuso, Jason, and Rahim Faiez. "Burns on Afghan Civilians Raise Concerns over White Phosphorus." *Arizona Republic,* 11 May 2009: A4.

Street, Paul. *Segregated Schools: Educational Apartheid in Post–Civil Rights America.* New York: Routledge, 2005.

Streitfeld, David, and Jack Healy. "Phoenix Leads the Way Down in Home Prices." *New York Times,* 29 April 2009. http://www.nytimes.com/2009/04/29/business/economy/29econ.html?_r=1&th&emc=th (accessed 29 April 2009).

Stricker, Frank. *Why America Lost the War on Poverty and How to Win It.* Chapel Hill: University of North Carolina Press, 2007.

"Study: 1 in 5 Young Adults Has Severe Personality Disorder." *Arizona Republic*, 2 December 2008: A4.

Summers, Graham. "The Real Crisis is Food: Beginning of the Bull for Agriculture." *Seeking Alpha*, 22 June 2009. http://seekingalpha.com/article/144675-the-real-crisis-is-food-beginning-of-the-bull-for-agriculture (accessed 22 June 2009).

Sutherland, Edwin Hardin. *White Collar Crime*. New York: Dryden Press, 1949.

Swarns, Rachel L. "2 Groups Compare Immigrant Detention Centers to Prisons." *New York Times*, 22 February 2007: A17.

Talan, J. "Scientists Link Anxiety to Specific Gene." *Arizona Republic*, 19 July 2002: A8.

Tanner, Lindsey. "Kids' Lower IQ Scores Linked to Prenatal Pollution." *Washington Post*, 20 July 2009: A1.

———. "Pediatric Group Urges Doctors to Check on Family's Exercise." *Arizona Republic*, 1 May 2006: A1, A7.

"The Land of Opportunity?" *New York Times*, 13 July 2007. http://www.nytimes.com/2007/07/13/opinion/13fri2.html (accessed 13 July 2007).

"The Social Security Act of 1935." The National Center Social Security Administration. n.d. http://www.nationalcenter.org/SocialSecurityAct.html (accessed 7 May 2008).

Thoma, Mark. "Goodbye, Homo Economicus." *RGE Monitor*, 26 March 2009. http://www.rgemonitor.com/financemarketsmonitor/256185/goodbye_homo_economicus (accessed 26 March 2009)

Thompson, Craig J. "Natural Health Discourses and the Therapeutic Production of Consumer Resistance." *The Sociological Quarterly*, 441 (2003): 81–107.

"Today's Youths Think Highly of Themselves." *Arizona Republic*, 19 November 2008: A15.

Tomes, Nancy. *The Gospel of Germs: Men, Women, and the Microbe in American Life*. Cambridge, MA: Harvard University Press, 1998.

Tomsho, Robert. "Study Confirms Antipsychotics Pose Heart Risk." *Wall Street Journal*, 15 January 2009: D1, D6.

Toppo, Greg. "For Many Kids, Lead Threat is Right in Their Own Homes." *USA Today*, 29 October 2007. http://www.usatoday.com/news/health/2007-10-28-lead-cover_N.htm (accessed 8 May 2008).

Torello, Michael W., and Frank H. Duffy. "Using Brain Electrical Activity Mapping to Diagnose Learning Disabilities." *Theory Into Practice* 24, no. 2 (Spring 1985): 95–99.

Tough, Paul. "What It Takes to Make a Student." *New York Times Magazine*, 26 November 2006. http://www.nytimes.com/2006/11/26/magazine/26tough.html?_r=1&oref=slogin (accessed 26 November 2006).

Tugend, Alina. "The (Possible) Perils of Being Thirsty while Being Green." *New York Times*, 5 January 2008. http://www.nytimes.com/2008/01/05/business/.../05shortcuts.html (accessed 5 January 2008).

Turse, Nick. *The Complex: How the Military Invades Our Everyday Lives*. New York: Metropolitan Books, 2008.

"Tranquilizing Drugs and Behavioral Disorders." *Science New Series* 124, no. 3215 (August 1956): 259.

Transit, Roxanna P., ed. *Disciplining the Child via the Discourse of the Professions*. Springfield, IL: Charles C. Thomas, 2004.

————. "The Ordering of Attention: The Discourse of Developmental Theory and ADD." In *Disciplining the Child via the Discourse of the Professions*, edited by Roxana Transit, 34–62. Springfield, IL: Charles C. Thomas, 2004.

Trent, J. W. *Inventing the Feeble Mind: A History of Mental Retardation in the United States*. Berkeley: University of California Press, 1994.

Trottman, Melanie, and Elizabeth Williamson. "Children's Product Industry Put in Regulatory Bind." *Wall Street Journal*, 18 June 2008: A3.

Twenge, Jean M. "Self-Entitlement of Grads Can Be a Curse." *Arizona Republic*, 30 May 2009: B5.

"U.S. Blocks Small Arms Controls." *BBC*, 10 July 2001. http://news.bbc.co.uk/2/hi/africa/1430077.stm (accessed 8 June 2008).

U.S. Department of Agriculture. "National School Lunch Program." 27 May 2009. http://www.fns.usda.gov/cnd/lunch/AboutLunch/ProgramHistory_5.htm#APPROVED (accessed 9 June 2009).

U.S. Department of Homeland Security. "Rightwing Extremism: Current Economic and Political Climate Fueling Resurgence in Radicalization and Recruitment" 7 April 2009. http://www.fas.org/irp/eprint/rightwing.pdf (accessed 3 May 2009).

U.S. Department of Labor. "The Employment Situation Summary June 2009." Bureau of Labor Statistics. U.S. Department of Labor, 2 July 2009. http://www.bls.gov/news.release/empsit.nr0.htm (accessed 7 July 2009).

————. "History of Federal Minimum Wage Rates under the Fair Labor Standards Act, 1938–2007." U.S. Department of Labor Employment Standards Administration, 22 September 2008. http://www.dol.gov/ESA/minwage/chart.htm (accessed 9 May 2009).

————. "Union Members Survey." Bureau of Labor Statistics. U.S. Department of Labor, 28 January 2009. http://www.bls.gov/news.release/union2.nr0.htm (accessed 20 February 2009).

U.S. Government Accounting Office. "Chemical Regulation: Comparison of U.S. and Recently Enacted European Union Approaches to Protect against the Risks of Toxic Chemicals GAO-07–825." 17 August 2007. http://www.gao.gov/docdblite/details.php?rptno=GAO-07–825 (accessed 9 September 2008).

Uchitelle, Louis. "Nafta Should Have Stopped Illegal Immigration, Right?" *New York Times*, 18 February 2007. http://www.nytimes.com/2007/02/18/weekinreview/18uchitelle.html (accessed 18 February 2007).

"Unemployment Rate Skyrockets." *Military.com*, 14 November 2005. http://www.military.com/NewsContent/0,13319,80320,00.html (accessed 9 October 2008).

United Nations. "Refugee Children: Guidelines and Care." UNHCR: The U.N. Refugee Agency, 1 January 1994. http://www.unhcr.org/cgi-bin/texis/vtx/search?page=search&docid=3b84c6c67&query=guidelines on refugee children (accessed 20 May 2009).

"U.S. 'Uses Incendiary Arms' in Iraq." *The BBC*, 9 November 2005. http://news.bbc.co.uk/2/hi/middle_east/4417024.stm (accessed 8 May 2009).

"VA to Increase Benefits for Mild Brain Trauma." *USA Today*, 23 September 2008. http://www.usatoday.com/news/military/2008-09-22-tbibenefits_N.htm (accessed 9 May 2009).

Vallianatos, Evaggelos. "From Watchdog to Lapdog: An Insider's History of the EPA." *Alternet*, 20 May 2009. http://www.alternet.org/story/140349/ (accessed 31 May 2009).

Valverde, M., and Mopas, M. "Insecurity and the Dream of Targeted Governance." In *Global Governmentality*, edited by W. Larner and W. Walters, 232–50. New York: Routledge, 2004.

Van Arsdol, D. Maurice D., Georges Sabagh, and Francesca Alexander. "Reality and the Perception of Environmental Hazards." *Journal of Health and Human Behavior* 5, no. 4 (Winter 1964): 144–53.

Vandell, Deborah Lowe, and Barbara Wolfe. "Child Care Quality: Does It Matter and Does It Need to Be Improved? Report Prepared for Office of the Assistant Secretary for Planning and Evaluation, U.S. Department of Health and Human Services." May 2000. http://aspe.hhs.gov/hsp/ccquality00/index.htm (accessed 9 March 2008)

Van Krieken, Robert. "Social Theory and Child Welfare." *Theory and Society* 15 (1986): 401–29.

Vanobbergen, Bruno. "Wanted: Real Children. About Innocence and Nostalgia in a Commodified Childhood." *Studies in Philosophy and Education* 23, nos. 2/3 (2004): 161–76.

Vasudevan, Ramaa. "Financialization: A Primer." *Dollars and Sense*, 2008. http://www.dollarsandsense.org/archives/2008/1108vasudevan.html (accessed 9 February 2009).

"Vatican Deplores Gaza Situation." *BBC News,* 8 January 2009. http://news.bbc.co.uk/2/hi/europe/7817019.stm (accessed 8 January 2009).

Vedantam, Shanker. "Climate Fears are Driving 'Ecomigration' across Globe." *Washington Post,* 23 February 2009: A1.

———. "Debate over Drugs for ADHD Reignites." *Washington Post,* 27 March 2009: A1.

———. "Researchers Find Stress, Depression Have Genetic Link." *Arizona Republic,* 18 July 2003: A10.

Vogel, Ann. "Who's Making Global Civil Society? Philanthropy and U.S. Empire in World Society." *British Journal of Sociology* 57 (2006): 635–55.

Wachs, Theodore D., and Pattiann Cucinotta. "The Effects of Enriched Neonatal Experiences upon Later Cognitive Functioning." *Developmental Psychology* 5, no. 3 (1971): 542.

Walk, Alexander. "The Pre-History of Child Psychiatry." *British Journal of Psychiatry* 110 (1964): 754–67.

Wallis, Claudia. "Does Watching TV Cause Autism?" *Time,* 20 October 2006. http://www.time.com/time/health/article/0,8599,1548682,00.html (accessed 5 November 2008).

Walsh, Mary Williams. "Actuaries Scrutinized on Pensions." *New York Times,* 21 May 2008. http://www.nytimes.com/2008/05/21/business/21pension.html (accessed 21 May 2008).

———. "Once Safe, Public Pensions Are Now Facing Cuts." *New York Times,* 6 November 2006. http://www.nytimes.com/2006/11/06/business/06pension.html?_r=1&th&emc=th&oref=slogin (accessed 6 November 2006).

Wang, Shirley S. "Mental-Health Drug Usage Rises." *Wall Street Journal,* 5 May 2009: D5.

———. "The War on Obesity Targets Toddlers." *Wall Street Journal,* 10 June 2008: D1–2.

Warner, Judith. "Living the Off-Label Life." *New York Times,* 27 December 2008. http://www.nytimes.com/2008/12/27/opinion/27warner.html (accessed 27 December 2008).

————. "Mommy Madness." *Newsweek*, 11 April 2005: 92.

Warren, Christian. *A Brush with Death: A Social History of Lead Poisoning*. Baltimore: Johns Hopkins University Press, 2001.

Warren, Elizabeth, and Amelia Warren Tyagi. *Two Income Trap: Why Middle-Class Mothers and Fathers are Going Broke*. New York: Basic Books, 2003.

Warrick, Joby. "Experts See Security Risks in Downturn." *Washington Post*, 15 November 2008: A1.

Washington, Harriet. A. *Medical Apartheid: The Dark History of Medical Experimentation on Black Americans from Colonial Times to the Present*. New York: Doubleday, 2006.

Weintraub, Sidney. "Confused and Mean-Spirited US Handling of Immigration Problems." *Issues in International Political Economy* 102 (2008): 1–2. http://www.csis.org/media/csis/pubs/issues200806.pdf (accessed 11 May 2009).

Weisman, Jonathan. "State's Budget Gaps Are Another Test for Washington." *Wall Street Journal*, 15 June 2009: A2.

Weiss, Gabrielle, John Werry, Klaus Minde, Virginia Douglas, and Donald Sykes. "Studies on the Hyperactive Child: V. The Effects of Dextroamphetamine and Chlorpromazine on Behavior and Intellectual Functioning." *Journal of Child Psychology and Psychiatry* 9, nos. 3–4 (1968): 145–56.

Weiss, M. "Abstinence-Only Sex Ed Finds Few Scientific Fans: Birth Control Taught in Shrinking Number of Schools, Study Says." *San Francisco Chronicle*, 11 February 2007: A17.

Wells, D. C. "Social Darwinism." *American Journal of Sociology* 12 (1907): 695–716.

Werry, John S. "Studies on the Hyperactive Child: IV An Empirical Analysis of the Minimal Brain Dysfunction Syndrome." *Archives of General Psychiatry* 19, no. 1 (1968): 9–16.

Werry, John S., Gabrielle Weiss, Virginal Douglas, and Judith Martin. "Studies on the Hyperactive Child: III. The Effect of Chlorpromazine upon Behavior and Learning Ability." *Journal of the American Academy of Child Psychiatry* 5, no. 2 (1966): 292–312.

Wertz, Richard W., and Dorothy C. Wertz. *Lying-In a History of Childbirth in America*. New Haven, CT: Yale University Press, 1989

Wessel, David. "Changing Attack: In Poverty Tactics, an Old Debate: Who is At Fault?" *Wall Street Journal*, 15 June 2006: A1, A10.

————. "Why Job Market Is Sagging in the Middle." *Wall Street Journal*, 11 October 2007: A2.

"When Head Start Is Too Late." *New York Times*, 12 August 1988: A26.

Whoriskey, Peter. "Out of Work and Challenged on Benefits, Too." *Washington Post*, 12 February 2009: A1.

Wilder, Douglas L. "To Save the Black Family, the Young Must Abstain." *Wall Street Journal*, 28 March 1991: A14.

Williamson, Elizabeth. "Some Americans Lack Food but USDA Won't Call Them Hungry." *Washington Post*, 16 November 2006: A1.

Wilson, Jamie. "US Admits Using White Phosphorus in Falluja." *Guardian*, 16 November 2005. http://www.guardian.co.uk/world/2005/nov/16/iraq.usa (accessed 9 May 2008).

Wilson, William Julius. *The Declining Significance of Race: Blacks and Changing American Institutions*. Chicago: University of Chicago Press, 1980.

Wines, Michael. "China's Leader Says He Is 'Worried' over U.S. Treasuries." *New York Times*, 14 March 2009. http://www.nytimes.com/2009/03/14/business/worldbusiness/14china.html?_r=1&th&emc=th (accessed 14 March 2009).

Winnicott, D. W. "Transitional Objects and Transitional Phenomena." In *Identity: A Reader*, edited by Paul DuGay, Jessical Evans, and Peter Redman, 150–62. London: Sage, 2000.

Winslow, C. E. A. "The Untilled Fields of Public Health." *Science* 51 (1920): 23–33.

Witte, Griff. "For Children of Gaza, Scars to Last a Lifetime." *Washington Post*, 26 January 2009: A5.

Wolff, Richard. "Capitalism Hits the Fan: A Marxian View." Google Videos, 9 October 2008. http://video.google.com/videoplay?docid=7382297202053077236 (accessed 10 January 2009).

Wollons, Roberta. Introduction. *Children at Risk in America: History, Concepts, and Public Policy*, edited by Roberta Wollons, ix-xxv. Albany: State University of New York Press, 1993.

"World Bank Finds More People Live in Steep Poverty." *New York Times*, 27 August 2008: A9.

World Bank. "Global Development Finance 2009: Charting a Global Recovery." World Bank Global Development Finance, June 2009. http://web.worldbank.org/WBSITE/EXTERNAL/EXTDEC/EXTDECPROSPECTS/EXTGDF/EXTGDF2009/0,,contentMDK:22218327~menuPK:5924239~pagePK:64168445~piPK:64168309~theSitePK:5924232,00.html (accessed *22 June 2009*).

———. *"World Development Indicators."* 2008. http://web.worldbank.org/WBSITE/EXTERNAL/DATASTATISTICS/0,,contentMDK:21725423~pagePK:64133150~piPK:64133175~theSitePK:239419,00.html (accessed 9 April 2009).

Worth, Robert F. "Hezbollah Shrine to Terrorist Suspect Enthralls Lebanese Children." *New York Times*, 3 September 2008. http://www.nytimes.com/2008/09/03/world/middleeast/03lebanon.html (accessed 3 September 2008).

Wright J. P., K. Dietrich, M. D. Ris, R. W. Hornung, and S. D. Wessel. "Association of Prenatal and Childhood Blood Lead Concentrations with Criminal Arrests in Early Adulthood." *PLoS Medicine* 5, no. 5 (2008): e101.

Wright, Robert. "The Biology of Violence." In *The Biological Basis of Human Behavior*, 2nd ed., edited by Robert W. Sussman, 257–65. Upper Saddle River, NJ: Prentice Hall, 1999.

Zelizer, Viviana A. *Pricing the Priceless Child: The Changing Social Value of Children*. New York: Basic Books, 1985.

Zoroya, Gregg. "VA Report: Male U.S. Veteran Suicides at Highest in 2006." *USA Today*, 8 September 2008. http://www.usatoday.com/news/military/2008-09-08-Vet-suicides_N.htm (accessed 9 November 2008).

Zuboff, Shoshana. *In the Age of the Smart Machine: The Future of Work and Power*. New York: Basic Books, 1988.

INDEX